43 iwg 120
Uy 32804

Ausgeschieden
im Jahr 2025

NUCLEAR CONDENSED MATTER PHYSICS

Günter Schatz studied physics at the Free University of Berlin and received his diploma in physics with experimental work done at the Hahn-Meitner-Institut, Berlin. From 1969 to 1976 he was assistant at the University Erlangen/Nürnberg (PhD in 1971). Since 1976 he has been professor of physics at the University of Konstanz. He spent two years (1973–1975) at the State University of New York in Stony Brook, USA (with a Max-Kade fellowship) and has visited this university several times since then for research stays. He has been several times to the Weizmann Institute in Israel for scientific work. His recent research is in the field of surface science with radioactive probes.

Alois Weidinger studied physics at the University of Mainz and the Free University of Berlin. He received the diploma in 1966 and a PhD in 1969 with experimental work done at the Hahn-Meitner-Institut Berlin. From 1969 to 1975 he was assistant at the University of Munich and from 1975 to 1987 research fellow at the University of Konstanz. Since 1988 he has been a scientist at the Hahn-Meitner-Institut Berlin. In 1966/67 he spent a year as a NATO fellow at the Yale University in the USA in the laboratory of Professor A. D. Bromley; in 1972/73 he was research fellow at the University of Paris-Sud in Orsay, France. His recent research is in the field of hydrogen in metals and in fullerene science.

John Gardner has been a faculty member at Oregon State University since 1973. He received a BA in Physics and Mathematics from Rice University in 1961, a MS (1963) and PhD (1966) from the University of Illinois. He was a faculty member at the University of Pennsylvania before going to Oregon State University. He has visited Germany several times—a postdoctoral year at the Technical University of Munich, 1966 and extended leaves at the Max Planck Institute for Solid State Physics, Stuttgart, the University of Stuttgart, and the University of Konstanz. He is a solid state experimental physicist employing several nuclear methods in his research.

NUCLEAR CONDENSED MATTER PHYSICS
Nuclear Methods and Applications

Günter Schatz
Department of Physics, University of Konstanz

Alois Weidinger
Solid State Physics Department, Hahn-Meitner-Institute Berlin

Translated by
John A. Gardner
Department of Physics, Oregon State University, USA

JOHN WILEY & SONS
Chichester · New York · Brisbane · Toronto · Singapore

© B. G. Teubner, Stuttgart, 1992:
G. Schatz/A. Weidinger, Nukleare Festkörperphysik, second edition.
This book is published by permission of B. G. Teubner and is the sole authorized English translation of the original German edition

Copyright © 1996 by John Wiley & Sons Ltd,
  Baffins Lane, Chichester,
  West Sussex PO19 1UD, England

Telephone: National   01243 779777
             International (+44) 1243 779777

All rights reserved.

No part of this book may be reproduced by any means, or transmitted, or translated into a machine language without the written permission of the publisher.

Other Wiley Editorial Offices

John Wiley & Sons, Inc., 605 Third Avenue,
New York, NY 10158-0012, USA

Jacaranda Wiley Ltd, 33 Park Road, Milton,
Queensland 4064, Australia

John Wiley & Sons (Canada) Ltd, 22 Worcester Road,
Rexdale, Ontario M9W 1L1, Canada

John Wiley & Sons (SEA) Pte Ltd, 37 Jalan Pemimpin #05-04,
Block B, Union Industrial Building, Singapore 2057

Library of Congress Cataloging-in-Publication Data
Schatz, Günter.
  [Nukleare Festkörperphysik. English]
  Nuclear condensed matter physics : nuclear methods and applications / Günter Schatz, Alois Weidinger ; translated by John A. Gardner. — 2nd ed.
    p. cm.
  Includes bibliographical references and index.
  ISBN 0-471-95479-9 (alk. paper)
  1. Nuclear matter. 2. Condensed matter. 3. Nuclear physics–Methodology. I. Weidinger, Alois. II. Title.
QC793.3.N8S313   1995
530.4'1'0724—dc20   95-4911
                                                        CIP
British Library Cataloguing in Publication Data

A catalogue record for this book is available from the British Library

ISBN 0 471 95479 9

Typeset in 10/12pt Times by Keytec Typesetting Ltd, Bridport, Dorset
Printed and bound in Great Britain by Biddles Ltd, Guildford, Surrey
This book is printed on acid-free paper responsibly manufactured from sustainable forestation, for which at least two trees are planted for each one used for paper production.

Contents

Preface ix

1 Introduction 1

2 Electromagnetic Properties and Nuclear Decay 3
   2.1 The Nuclear Magnetic Dipole Moment 3
   2.2 The Electric Quadrupole Moment 7
   2.3 Gamma-Decay of the Nucleus 10
   2.4 Detection of γ-Radiation 15

3 Hyperfine Interactions 21
   3.1 Magnetic Interaction 21
   3.2 Electric Interaction 25

4 Mössbauer Effect 33
   4.1 Principles 33
   4.2 The Debye–Waller Factor 36
   4.3 Mössbauer Sources and Experimental Apparatus 43
   4.4 Isomer Shift 48
   4.5 Electric Quadrupole Interation 53
   4.6 Magnetic Dipole Interaction 56
   4.7 Quadratic Doppler Effect 60

5 Perturbed γ–γ Angular Correlation (PAC) 63
   5.1 Theory of Unperturbed γ–γ Angular Correlations 63
   5.2 Theory of Perturbed γ–γ Angular Correlations 71
   5.3 Calculation of the Perturbation Factor for Special Cases 73
   5.4 PAC Sources and Experimental Apparatus 78
   5.5 Electric Field Gradients in Non-Cubic Metals 86

	5.6	Atomic Defects in Metals	89
	5.7	Adsorbate Sites on Surfaces	92
	5.8	Internal Magnetic Fields in Ferromagnetic Materials	94
	5.9	Integral Perturbed Angular Correlation (IPAC) and Transient Magnetic Fields in Ferromagnets	95

6 Nuclear Magnetic Resonance (NMR) — 101
- 6.1 Principles — 101
- 6.2 Classical Treatment of NMR (Bloch Equations) — 104
- 6.3 Experimental Methods — 110
- 6.4 Chemical Shift — 118
- 6.5 Knight Shift in Metals — 123
- 6.6 Spin–Lattice Relaxation — 128
- 6.7 NMR with Radioactive Nuclei and Self-Diffusion in Metals — 134

7 Nuclear Orientation (NO) — 139
- 7.1 Principles — 139
- 7.2 Experimental Apparatus — 141
- 7.3 Hyperfine Fields — 146
- 7.4 Spin–Lattice Relaxation at Low Temperatures — 146

8 Muon Spin Rotation (μSR) — 151
- 8.1 Principles — 151
- 8.2 Experimental Methods — 153
- 8.3 Internal B Fields in Magnetic Materials — 157
- 8.4 Diffusion of Positive Muons — 162
- 8.5 Muonium in Semiconductors — 174

9 Positron Annihilation — 181
- 9.1 Principles — 181
- 9.2 Positron Sources and Experimental Apparatus — 183
- 9.3 Angular Correlation of Annihilation Radiation (ACAR) and Fermi Momentum of Conduction Electrons in Metals — 186
- 9.4 Positron Lifetime and Lattice Defects in Metals — 189

10 Neutron Scattering — 195
- 10.1 Properties of Neutrons and Production of Neutron Beams — 196
- 10.2 Detection of Neutrons — 199
- 10.3 Theory of Neutron Scattering — 201
- 10.4 Elastic Neutron Scattering — 207
- 10.5 Quasi-Elastic Neutron Scattering — 212
- 10.6 Inelastic Neutron Scattering — 219

11	**Ion Beam Analysis**	223
	11.1 Rutherford Backscattering (RBS)	224
	11.2 Channeling	243
	11.3 Nuclear Reaction Analysis (NRA)	251

Appendix 263
    A.1 Clebsch–Gordan Coefficients and $3j$-Symbols 263
    A.2 Spherical Tensors 265
    A.3 Wigner–Eckart Theorem 267

Bibliography of Advanced Topics 269

References 271

Index 275

Preface

The authors have been teaching nuclear condensed matter physics for a number of years at the University of Konstanz and have found no satisfactory comprehensive introductory textbook. The present book is intended to overcome this deficiency for an increasingly important field.

Nuclear experimental techniques and their application to condensed matter research are described here. The intention is that the book be used as a text for courses on nuclear condensed matter or applied nuclear physics, as a resource book for seminars and laboratory courses, and finally as introductory literature for researchers in any of the fields covered.

The first edition was widely accepted by faculty and students in Germany, but it was criticized because it did not cover ion beam analysis. A chapter on ion beam analysis was added in the second edition. This English edition is a translation of the German second edition.

For their support and encouragement during preparation of this book, the authors wish to thank their colleagues at the University of Konstanz. In particular, we thank Professor E. Recknagel for his strong encouragement. For criticism and suggestions on specific topics we owe thanks to colleagues in many universities. We are also grateful to the many students who have critically read through the book, corrected errors and helped clarify difficult passages.

Günter Schatz
Alois Weidinger

Konstanz, February 1994

Translator's note: For the most part, this is a sentence-by-sentence translation and, I hope, captures the informal nature that has made this book so popular in Germany. A few minor errors have been corrected, and a small number of notational changes made to conform with practices in English-speaking countries.

I wish to thank Professors Günter Schatz and Ekkehard Recknagel for their hospitality during a sabbatical visit during which this translation was done. I also thank Professor Alois Weidinger for his patient assistance in the translation process. My German is far too weak to have done this without his help.

I am grateful to the Alexander von Humboldt Foundation for a Humboldt Fellowship during my sabbatical year. The authors and I thank Mr Mark Preddy for converting our final notes into the L^{A}T$_E$X file from which this book was published.

John Gardner

Konstanz, June 1994

1
Introduction

Nuclear condensed matter physics incorporates elements of both nuclear and condensed matter physics. The contribution of nuclear physics lies mainly in the methodology, i.e. in the experimental apparatus and the effects observed. In most cases, detection of particle and γ-radiation plays an important role. For these purposes, one uses instruments and methods developed for nuclear physics.

On the other hand, the subjects investigated arise from condensed matter physics. In this respect, nuclear condensed matter physics is purely a field of condensed matter physics. As in any new field, the experimental methods, e.g. the nuclear aspects, were originally of primary interest. In the intervening time, the methodological problems have been largely solved. That has the consequence that condensed matter physics, which is the primary research subject, has gained importance.

The term 'nuclear condensed matter physics' is not very precise. It needs further explanation. The following can be used as a shorthand description: In nuclear condensed matter physics, one investigates condensed matter properties with nuclear methods. Nuclear experimental methods involve the detection of particles (electrons, protons, neutrons, α-particles, etc.) or γ-radiation from nuclear decays or nuclear reactions. In a broader sense, elementary particles (e.g. muons and positrons) interacting with condensed matter are included here as part of the field of nuclear condensed matter physics. Nuclear magnetic resonance is also included, because of its many features in common with other nuclear methods, and because nuclei are used for experimental observation even though no particle or γ-radiation is detected.

In addition to these methods, techniques employing energetic particles from accelerators are included in the field of nuclear condensed matter physics. This subfield is known as ion beam analysis. We will discuss three examples (Rutherford backscattering, channeling, and nuclear reaction analysis) from this extraordinarily broad field.

2 Introduction

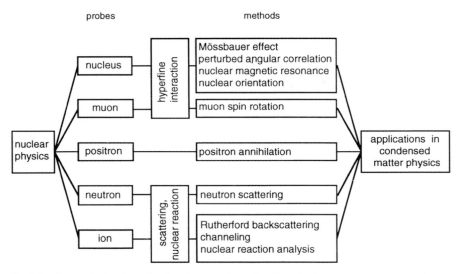

Fig. 1.1 Conceptual schematic of nuclear condensed matter physics.

Tracer methods, used for the study of the spatial and temporal distribution of radioactive atoms in solids, could, in principle, be a subject for this book. We have left this subject out and refer the reader to detailed review articles (Askill, 1970).

Fig. 1.1 elucidates schematically the relationship between nuclear and condensed matter physics in the context of nuclear condensed matter physics. Starting from nuclear physics and the properties of individual particles (nuclei, muons, positrons, neutrons, and ions), their interactions with condensed matter leads to the different experimental methods, and finally to their applications to the study of matter.

2
Electromagnetic Properties and Nuclear Decay

The first topics to be treated in this book are those that involve nuclei as probes in condensed matter. In order that these nuclei be useful probes, it is necessary that a measurable interaction exists between the nuclei and the condensed matter. Since these interactions are electromagnetic, we will discuss the electric and magnetic properties of the probes, i.e. the nuclear magnetic dipole moment μ and the nuclear electric quadrupole moment Q. Information about the interaction of μ and Q with magnetic and electric fields is often observed via emission of γ-radiation. Therefore, in this chapter, we will also describe the characteristic γ-ray decay properties of nuclei.

2.1 The Nuclear Magnetic Dipole Moment

The magnetic moment μ of a charged body with angular moment I is given classically by

$$\mu = \gamma I \quad (2.1)$$

where γ is called the gyromagnetic ratio. In quantum mechanics, μ and I are operators, and one defines the magnetic moment as the value of the z component of μ in the state $|I, M\rangle$ with $M = I$ (I, M are quantum numbers of total angular momentum and the z component respectively)

$$\mu := \langle I, M = I | \mu_z | I, M = I \rangle \quad (2.2)$$

This leads to

$$\mu = \gamma \hbar I \quad (2.3)$$

4 Electromagnetic Properties and Nuclear Decay

The gyromagnetic ratio of nuclei is often expressed by the dimensionless g-factor defined by

$$\gamma = g\frac{\mu_N}{\hbar} \qquad (2.4)$$

with nuclear magneton μ_N

$$\mu_N = \frac{e\hbar}{2m_p} = 5.05 \times 10^{-27} \quad \text{A m}^2 \qquad (2.5)$$

where m_p is the proton mass. For nuclei with angular momentum L the g-factor can easily be calculated classically to be

$$g_L(\text{proton}) = 1 \qquad g_L(\text{neutron}) = 0 \qquad (2.6)$$

For free nucleons the experimental g-factor associated with the spin angular momentum S is

$$g_S(\text{proton}) = 5.59 \qquad g_S(\text{neutron}) = -3.83 \qquad (2.7)$$

For pure Dirac particles one would expect $g_S = 2$ for protons and $g_S = 0$ for neutrons. The strong deviation of measured values from the Dirac theory indicates that protons and neutrons are composite particles. In the quark theory, protons and neutrons are each composed of three quarks: p = (uud), n = (udd) where u and d are the up and down quarks respectively. Using the generalized Landé formula (see Eq. (2.11)) one can calculate the magnetic moment of nucleons within the quark model. One obtains values that are approximately correct.

If the magnetic moment of the nucleus is due largely to a single nucleon outside a closed nuclear shell, one can easily derive the g-factor. The wave function is

$$\Psi_I = \Phi_{ILS}\Omega_0 \qquad (2.8)$$

where Ω_0 is the inert core with I and μ both zero. For the magnetic moment μ and angular momentum I of the outer nucleon, the following operator equations hold:

$$\mu = g_L L + g_S S \quad \text{with } I = L + S \quad \text{and } S = 1/2 \qquad (2.9)$$

where μ is given in units of μ_N and I, L, and S in units of $\hbar$.

Fig. 2.1 shows the coupling scheme of the angular momenta. For arbitrary

The Nuclear Magnetic Dipole Moment

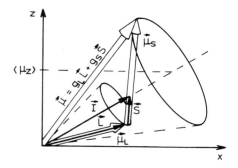

Fig. 2.1 Coupling scheme for angular momenta and magnetic moments. Only the component of μ parallel to I is observable. Perpendicular components average to zero.

angular momenta I_1 and I_2, one can derive from this coupling scheme the generalized Landé formula. With

$$\mu = \mu_1 + \mu_2 \qquad \mu = g(I)I$$
$$\mu_1 = g(I_1)I_1 \qquad (2.10)$$
$$I = I_1 + I_2 \qquad \mu_2 = g(I_2)I_2$$

one obtains

$$g(I) = \frac{1}{2I(I+1)}\{[I(I+1) + I_1(I_1+1) - I_2(I_2+1)]g(I_1)$$
$$+ [I(I+1) + I_2(I_2+1) - I_1(I_1+1)]g(I_2)\} \qquad (2.11)$$

Using the generalized Landé formula, the g-factor for a single particle is found to be

$$g = g_L \pm \frac{g_S - g_L}{2L+1} \qquad (I = L \pm 1/2) \qquad (2.12)$$

Using the g_S values of free nucleons (Eq. (2.7)), along with $g_L = 1$ for protons and $g_L = 0$ for neutrons, one obtains the Schmidt values. Dirac values are obtained from Eq. (2.12) using $g_S = 2$ for the proton and $g_S = 0$ for the neutron. As can be seen in Fig. 2.2, almost all experimental values lie between the Schmidt and Dirac values. Thus, with few exceptions, the g_S values for protons in nuclei are between 2 and 5.59, and those for neutrons between -3.83 and 0.

The deviation of g_S from the free nucleon value can be associated with a

6 Electromagnetic Properties and Nuclear Decay

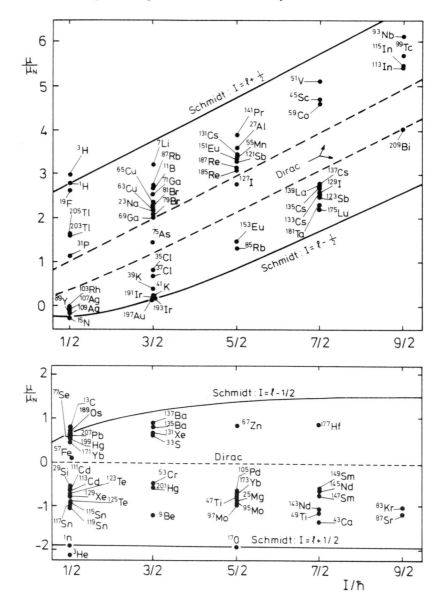

Fig. 2.2 Experimental ground-state magnetic moments of nuclei with an unpaired proton (upper) or neutron (lower). The theoretical values obtained using the Schmidt model (solid line) and the Dirac model (dashed line) are also shown. Adapted from Mayer-Kuckuk (1984); experimental values from Lederer and Shirley (1978).

polarization of the nuclear core by the orbiting outer nucleon (core polarization).

2.2 The Electric Quadrupole Moment

The classical quadrupole moment of a charge distribution $\rho(r)$ is given by

$$Q = \frac{1}{e}\int (3z^2 - r^2)\rho(r)\,d^3r = \frac{1}{e}\sqrt{\frac{16\pi}{5}}\int r^2 Y_2^0 \rho(r)\,d^3r \qquad (2.13)$$

The quantum mechanical definition of Q is the expectation value of the quadrupole operator $\sqrt{(16\pi/5)}\, Zr^2 Y_2^0$ in the state $|I, M = I\rangle$

$$Q := \sqrt{\frac{16\pi}{5}} \langle I, M = I | Zr^2 Y_2^0 | I, M = I \rangle \qquad (2.14)$$

where Y_2^0 is the spherical harmonic with $l = 2$, $m = 0$, and Z is the atomic number. The quantity $Q_{20} = zr^2 Y_2^0$ is a tensor operator. Using the Wigner–Eckart theorem (see Appendix A.3) one obtains

$$\langle I, M | Q_{20} | I, M \rangle = (-)^{I-M} \begin{pmatrix} I & 2 & I \\ -M & 0 & M \end{pmatrix} \langle I \| Q_2 \| I \rangle \qquad (2.15)$$

For $I < 1$, the 3j-symbol is zero, and therefore $Q = 0$, since I and 2 cannot be coupled to give a vector of length I (see Appendix A.1).

The electric quadrupole moment is a measure of the deviation of the nuclear charge density from spherical. As can be seen immediately from Eq. (2.13), the quadrupole moment of a spherically symmetric charge distribution is zero. Q has dimensions of length squared, or area. Normally, Q is expressed in units of barns where

$$1\text{ barn} = 100\text{ fm}^2 = 10^{-28}\text{ m}^2 \qquad (2.16)$$

Some experimental quadrupole moments are shown in Fig. 2.3.

The quadrupole moments can be calculated using simple nuclear models. In the single-particle model, the outer nucleon largely determines the electromagnetic properties. For the core, we have $I = 0$ and therefore $Q(\text{core}) = 0$, so that only the quadrupole moment due to the outer nucleon must be calculated. For a single proton, one obtains (Kamke, 1979)

8   Electromagnetic Properties and Nuclear Decay

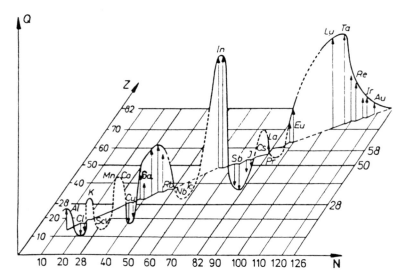

Fig. 2.3 Observed values of the nuclear quadrupole moment as a function of the atomic number Z and neutron number N (all values up to In are multiplied by 10). Adapted from Kopfermann (1956).

$$Q(\text{proton}) = -\frac{2I-1}{2(I+1)}\langle r^2 \rangle \tag{2.17}$$

where $\langle r^2 \rangle = \int \psi^* r^2 \psi \, d^3 r$ represents the average square radius of the proton orbit.

The quadrupole moment of a nucleus with a single proton orbiting around a closed shell is always negative. One can understand this result qualitatively, since for $M = I$, the proton wave function has its largest density near the equatorial plane, so $3z^2 - r^2$ is smaller than zero on average (see Fig. 2.4). In general

(a) Q is negative for disk-like charge distribution;
(b) Q is positive for cigar-shaped charge distribution.

For a neutron orbiting a core, these simple considerations would suggest $Q = 0$. However, the orbiting neutron causes a movement of the charged core about the common center of mass. A core with nuclear charge Ze orbits the center of mass at a distance of R/A where R is the nuclear radius, and A is the number of nucleons in the core. The mean square radius is $(R/A)^2$, resulting in a quadrupole moment

$$Q(\text{neutron}) = (Z/A^2) Q(\text{proton}) \tag{2.18}$$

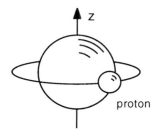

Fig. 2.4 Schematic explanation of the quadrupole moment for a single proton orbiting around a closed shell.

In addition, one finds that the orbiting nucleon polarizes the core by nuclear forces (core polarization) which contributes strongly to the quadrupole moment. One accounts for this core polarization effect qualitatively by using an effective charge for the orbiting nucleon, leading to

$$Q = \frac{1}{e}\langle I, M = I | e^{\text{eff}}(3z^2 - r^2) | I, M = I \rangle = -\frac{e^{\text{eff}}}{e}\frac{2I-1}{2(I+1)}\langle r^2 \rangle \quad (2.19)$$

Typical values of the effective charges are

$$\frac{e^{\text{eff}}}{e}(\text{proton}) = 1 + e^{\text{pol}} \approx 1.2 \qquad \frac{e^{\text{eff}}}{e}(\text{neutron}) = e^{\text{pol}} \approx 0.6 \quad (2.20)$$

This simple single-particle model always leads to negative quadrupole moments, but both positive and negative quadrupole moments are found experimentally. The observation of positive values indicates that many nucleons can contribute to the quadrupole moment.

Effects of contributions by many nucleons may be taken approximately into account if we describe the nucleus as a homogeneously charged ellipsoid with charge Ze and semi-axes a and b (b in the z direction). In this case Eq. (2.13) gives for the quadrupole moment

$$Q = \tfrac{2}{5}Z(b^2 - a^2) = \tfrac{4}{5}Z\bar{R}^2\delta \quad (2.21)$$

with $\bar{R} = (a+b)/2$ and the deformation parameter $\delta = (b-a)/\bar{R}$. Typical values of the deformation δ are of order 10%. Extremely large deformations ($\delta \approx 100\%$) are found for fission isomers (Metag et al., 1980). This is expected, since during fission the nucleus goes through an intermediate state having a waist before separation of the two parts. Fission isomers are metastable nuclei formed prior to fission.

2.3 Gamma-Decay of the Nucleus

Quantum levels of nuclei are characterized by well-defined total angular momentum I and parity π (we ignore slight parity mixing caused by the weak interaction). In the transition from a higher to lower energy level gamma-rays are often emitted. The following conservation relations hold:

$$\begin{aligned} \text{energy} \quad & E_i = E_f + \hbar\omega \\ \text{angular momentum} \quad & \mathbf{I}_i = \mathbf{I}_f + \mathbf{l} \\ \text{parity} \quad & \pi_i = \pi_f \pi \end{aligned} \tag{2.22}$$

During the transition from the initial state (I_i, M_i, π_i) to the final state (I_f, M_f, π_f), a photon with quantum numbers (l, m, π) is emitted. The emission of a photon is equivalent to the creation of an electromagnetic wave with energy $\hbar\omega$ (see Fig. 2.5). The conservation of angular momentum and parity requires that the emitted wave, like the nuclear states, has well-defined angular momentum and parity. We therefore look for solutions to the Maxwell equations expressed in the form of multipole radiation where angular momentum and parity are explicit parameters.

In free space the Maxwell equations are

$$\nabla \times \mathbf{E} = -\frac{\partial \mathbf{B}}{\partial t} \qquad \nabla \times \mathbf{B} = \frac{1}{c^2}\frac{\partial \mathbf{E}}{\partial t} \tag{2.23}$$

$$\nabla \cdot \mathbf{E} = 0 \qquad \nabla \cdot \mathbf{B} = 0$$

where $\mathbf{E}$ is the electric field, $\mathbf{B}$ is the magnetic flux density, and c is the velocity of light in free space. The solutions of the Maxwell equations as multipole fields (Jackson, 1962) have the following form (unnormalized and without the time dependence, $\exp(-i\omega t)$):

$$\begin{aligned} \mathbf{B}_l^m = f_l(kr)\mathbf{L}Y_l^m(\theta,\phi) \qquad & \mathbf{E}_l^m = i\frac{c}{k}\nabla \times \mathbf{B}_l^m \qquad & \text{(E)} \\ \mathbf{E}_l^m = f_l(kr)\mathbf{L}Y_l^m(\theta,\phi) \qquad & \mathbf{B}_m^l = -i\frac{1}{kc}\nabla \times \mathbf{E}_l^m \qquad & \text{(M)} \end{aligned} \tag{2.24}$$

where the $f_l(kr)$ depend on r but not on angle. They are proportional to spherical Bessel functions. $Y_l^m(\theta,\phi)$ are spherical harmonics and $\mathbf{L}$ is the angular momentum operator, $\mathbf{L} = -i\hbar(\mathbf{r} \times \nabla)$. Sometimes one uses the vector spherical harmonics

Gamma-Decay of the Nucleus 11

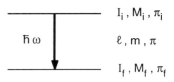

Fig. 2.5 Schematic representation of a γ transition between an initial state i and a final state f.

$$X_l^m(\theta, \phi) = \frac{1}{\sqrt{l(l+1)}} L Y_l^m(\theta, \phi) \qquad (2.25)$$

The solutions (Eq. (2.24)) are the desired multipole fields:

(a) electrical 2^l-poles: equations (E);
(b) magnetic 2^l-poles: equations (M).

These are the radiation fields of oscillating 2^l-poles.

So far we have considered only classical electromagnetic radiation. In quantum electrodynamics, the γ quantum corresponding to the multipole radiation l has an angular momentum with magnitude $\sqrt{l(l+1)}\hbar$ and z component $m\hbar$. In addition, the multipole fields have a definite parity:

$$\pi = (-1)^l \text{ for E radiation} \qquad \pi = (-1)^{l+1} \text{ for M radiation} \qquad (2.26)$$

Using the angular momentum selection rules

$$l = I_i + I_f, I_i + I_f - 1, \ldots, |I_i - I_f| \qquad (2.27)$$

one obtains the multipole transitions, which are summarized in Table 2.1. Only the transitions of lowest order are shown.

Table 2.1 Allowed multipole transitions for given angular momentum and parity change.

Angular momentum change ΔI		0 no 0 → 0	1	2	3
Parity change	yes	E1	E1	M2	E3
		(M2)	(M2)	E3	(M4)
	no	M1	M1	E2	M3
		E2	E2	(M3)	E4

12 Electromagnetic Properties and Nuclear Decay

For a given order, the magnetic transitions are usually much smaller than the electric transitions, so $M(l + 1)$ can be neglected compared to $E(l)$, but $M(l)$ is often of the same order of magnitude as $E(l + 1)$.

Angular radiation pattern

The angular distribution of emitted γ-rays can be calculated from the solutions to the Maxwell equations. We begin by noting that the energy flux density is the Poynting vector

$$S = \frac{1}{\mu_0}(E \times B) \qquad (2.28)$$

Since the source–detector distances are much larger than the (nuclear scale) source dimensions, the far-field solutions for E and B can be used. One has

$$\varepsilon_0|E|^2 = \frac{1}{\mu_0}|B|^2 \quad \text{i.e. } |E| = c|B| \qquad (2.29)$$

and E, B, and r are mutually perpendicular. In this far-field region,

$$|S| = c\varepsilon_0|E|^2 = \frac{c}{\mu_0}|B|^2 \qquad (2.30)$$

Using the explicit solutions (Eq. (2.24)) one obtains

$$\begin{aligned} |S| = c\varepsilon_0|E|^2 \propto |LY_l^m|^2 & \quad \text{for } Ml \text{ radiation} \\ |S| = \frac{c}{\mu_0}|B|^2 \propto |LY_l^m|^2 & \quad \text{for } El \text{ radiation} \end{aligned} \qquad (2.31)$$

One sees that the E and M radiation of the same multipole order have the same angular pattern. Thus one cannot distinguish them on the basis of the angular dependence of the radiation. To distinguish El from Ml radiation, a polarization measurement is necessary. To find the angular distribution explicitly, we must compute $|LY_l^m|^2$. It is convenient to write L as

$$\begin{aligned} L_+ &= L_x + iL_y & L_x &= \tfrac{1}{2}(L_+ + L_-) \\ L_- &= L_x - iL_y & \text{i.e.} \quad L_y &= \frac{1}{2i}(L_+ - L_-) \\ L_z &= L_z & L_z &= L_z \end{aligned} \qquad (2.32)$$

The components of L operating on Y_l^m are (Lindner, 1984)

$$L_+ Y_l^m = \hbar\sqrt{(l-m)(l+m+1)}\, Y_l^{m+1}$$
$$L_- Y_l^m = \hbar\sqrt{(l+m)(l-m+1)}\, Y_l^{m-1} \qquad (2.33)$$
$$L_z Y_l^m = \hbar m Y_l^m$$

from which we obtain

$$\begin{aligned}
|LY_l^m|^2 &= |L_x Y_l^m|^2 + |L_y Y_l^m|^2 + |L_z Y_l^m|^2 \\
&= \tfrac{1}{2}|L_+ Y_l^m|^2 + \tfrac{1}{2}|L_- Y_l^m|^2 + |L_z Y_l^m|^2 \\
&= \frac{\hbar^2}{2}(l-m)(l+m+1)|Y_l^{m+1}|^2 \qquad (2.34) \\
&\quad + \frac{\hbar^2}{2}(l+m)(l-m+1)|Y_l^{m-1}|^2 \\
&\quad + \hbar^2 m^2 |Y_l^m|^2
\end{aligned}$$

Using the relation

$$|Y_l^m(\theta,\phi)|^2 = \sum_k \frac{2l+1}{4\pi}(2k+1)\begin{pmatrix} l & l & k \\ m & -m & 0 \end{pmatrix}\begin{pmatrix} l & l & k \\ 0 & 0 & 0 \end{pmatrix} P_k(\cos\theta)$$

(2.35)

where $P_k(\cos\theta)$ are the Legendre polynomials, one can calculate the normalized angular distributions

$$F_{lm}(\theta) = \frac{|LY_l^m|^2}{\sum_m |LY_l^m|^2} \qquad (2.36)$$

From the properties of the 3j-symbols (Appendix A.1), the following conditions must hold:

$$k \leq 2l \text{ and } k \text{ even since } \begin{pmatrix} l & l & k \\ 0 & 0 & 0 \end{pmatrix} = 0 \text{ for } k \text{ odd} \qquad (2.37)$$

In Table 2.2 the angular distribution functions $F_{lm}(\theta)$ for dipole and quadrupole radiation are presented. They are shown schematically in Fig. 2.6.

Electromagnetic Properties and Nuclear Decay

Table 2.2 Angular distribution functions $F_{lm}(\theta)$ for dipole and quadrupole radiation.

	$m = 0$	$m = \pm 1$	$m = \pm 2$
$l = 1$ (dipole)	$\frac{1}{2}\sin^2\theta$	$\frac{1}{4}(1 + \cos^2\theta)$	—
$l = 2$ (quadrupole)	$\frac{3}{2}\sin^2\theta\cos^2\theta$	$\frac{1}{4}(1 - 3\cos^2\theta + 4\cos^4\theta)$	$\frac{1}{4}(1 - \cos^4\theta)$

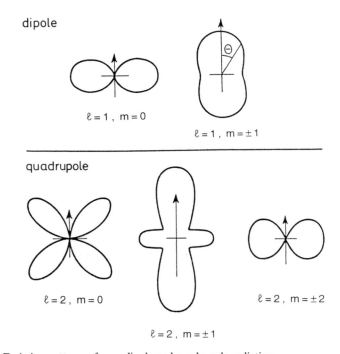

Fig. 2.6 Emission patterns of pure dipole and quadrupole radiation.

The functions $F_{lm}(\theta)$ possess some interesting properties:

(a) $\sum_m F_{lm}(\theta) = 1$ isotropy for the sum over all m
(b) $F_{lm}(\theta) = F_{l-m}(\theta)$ symmetry in m
(c) $F_{lm}(\theta) = F_{lm}(\pi - \theta)$ mirror symmetry in the xy plane
(d) $F_{lm}(\theta = 0°) = 0$ for $m \neq \pm 1$

2.4 Detection of γ-Radiation

Two basic interaction processes between γ-rays and matter permit γ-rays to be detected: the photoeffect and the Compton effect. Pair production does not play a major role in detecting the low-energy γ-rays normally of importance to nuclear condensed matter research.

2.4.1 Photoeffect

In the photoeffect, the γ-ray is absorbed and transfers its entire energy to a single electron. The kinetic energy E_e of the emitted electron is given by

$$E_e = E_\gamma - E_B \tag{2.38}$$

where E_B is the electron binding energy. The atom from which the electron is emitted is left in an excited state and relaxes by emitting X-rays or Auger electrons. Normally the energies of the emitted electron, Auger electrons, and X-rays are all deposited in the scintillator material, and therefore one obtains a signal proportional to the total γ-ray energy. The γ-ray photoelectric absorption cross section depends strongly on the atomic number Z and the γ-ray energy E_γ. The following proportionality is found to be approximately valid:

$$\sigma_{ph} \propto E_\gamma^{-7/2} Z^5 \tag{2.39}$$

We see that photoelectric absorption is strong for:

(a) small photon energies (light, X-rays and low-energy γ-rays);
(b) materials with high atomic number Z (e.g. iodine and lead).

2.4.2 Compton Effect

The Compton effect is the elastic scattering of γ-rays by free electrons. In general, electrons in solids are not free, but, at least for outer electrons, the binding is weak and can be neglected.

From the kinematics of the two-body collision (Fig. 2.7), one obtains for the energy of the scattered electron

$$E_e = E_\gamma \left[1 - \frac{1}{1 + (E_\gamma/m_e c^2)(1 - \cos\theta)} \right] \tag{2.40}$$

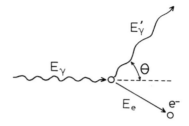

Fig. 2.7 Kinematics of the Compton effect. E_γ and E'_γ are the photon energies before and after the collision.

The scattered electrons have energies between $E_e = 0$ for $\theta = 0°$ and $E_e = E_{max}$ for $\theta = 180°$, where E_{max} is given by

$$E_{max} = E_\gamma \frac{1}{1 + m_e c^2/(2E_\gamma)} \qquad (2.41)$$

The energy spectrum of the scattered electrons can be derived using the Klein–Nishina formula (Klein and Nishina, 1929). This distribution is shown for two different γ-ray energies in Fig. 2.8.

The Compton cross section is proportional to the density of electrons and

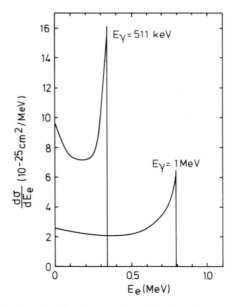

Fig. 2.8 Energy distributions for Compton-scattered electrons for two different γ-ray energies.

inversely proportional to the γ-ray energy E_γ. Since the electron density is approximately equal to the atomic number Z, one obtains

$$\sigma_C \propto E_\gamma^{-1} Z \qquad (2.42)$$

One sees that σ_C is less strongly dependent on E_γ or Z than is σ_{ph}. The Compton effect is most significant for γ-rays having moderate energies, i.e. between 100 keV and 1 MeV, and for materials with low Z.

The reduction in the intensity of γ-rays in passing through a material of thickness d is given by

$$I/I_0 = \exp(-\mu d) \qquad (2.43)$$

where μ is the absorption coefficient and I/I_0 is the ratio of the transmitted to the incident γ-ray intensity. The absorption coefficients for the different processes are displayed for NaI and Ge in Fig. 2.9. NaI and Ge are two important γ-ray detector materials.

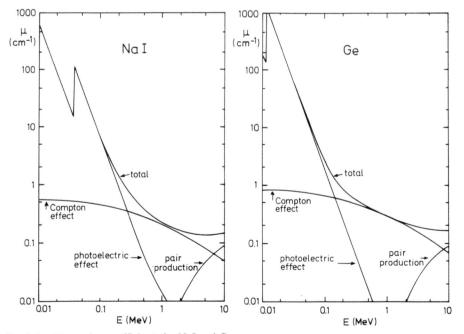

Fig. 2.9 Absorption coefficients for NaI and Ge.

2.4.3 Detectors

The electrons from the Compton and photoeffect lose their energy in the scintillator or semiconductor material and produce a signal that can be detected.

The schematic construction of a scintillation detector is shown in Fig. 2.10. The following processes occur during detection of a γ-ray:

(a) The primary electron produced by Compton or photoeffect loses its energy causing an ionization (or excitation) of the scintillator atoms; the number of ionized atoms is proportional to E_e.
(b) The ionized or excited atoms relax by emission of light.
(c) The light is absorbed by the photocathode of a photomultiplier tube, causing electrons to be emitted.
(d) The electron avalanche is amplified by dynodes of the photomultiplier tube.

Common scintillator materials

NaI(Tl) scintillator Because of the iodine, this scintillator has a large photoeffect efficiency. The photoelectrons are trapped by the Tl dopants, which then relax by emission of a photon having energy below the NaI bandgap energy. These photons are detected in the photomultiplier tube. The energy resolution of NaI is moderate ($\Delta E \approx 50$ keV for $E_\gamma = 1$ MeV) and has moderate time resolution ($\Delta t \approx 2$ ns). A typical energy spectrum is shown in Fig. 2.11.

BaF$_2$ scintillator The detection efficiency is better but the energy resolution somewhat worse than that of NaI(Tl). The main advantage of this detector is that a fraction of the scintillation light is emitted very quickly after the γ-ray is absorbed. This fast component permits very good time resolution ($\Delta t \approx 300$ ps for 511 keV γ-rays). The good time resolution and usually

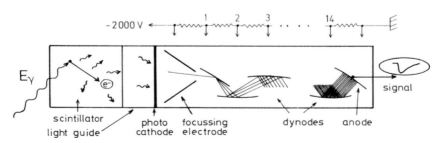

Fig. 2.10 Schematic construction of a scintillation detector.

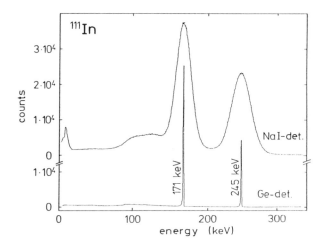

Fig. 2.11 Energy spectrum of a ^{111}In source taken with a NaI(Tl) detector (upper spectrum) and an i-Ge detector (lower spectrum).

adequate energy resolution make this an increasingly common scintillator. Since the fast component is in the ultraviolet, a quartz window is necessary for the photomultiplier tube.

Plastic scintillator This detector is used to attain high time resolution (typically $\Delta t \approx 200$ ps). However, these materials do not contain heavy elements, so the primary γ-ray interaction process is Compton scattering, and energy resolution is poor.

Semiconductor detectors

In addition to scintillator materials, semiconductors also play an important role in γ-ray detection. The initial photoelectron and Compton processes occur in these materials as in scintillators, but the electrons and holes excited in the absorption process are detected electrically. It is critical that the electrical conductivity of the semiconductor be vanishingly small so that most of the charge carriers are those created by the γ-ray absorption. Both silicon and germanium are suitable materials, but silicon is useful only for small γ-ray energies because of its small atomic number. We discuss only the Ge detector here.

Low-conductivity Ge can be obtained in two ways:

Intrinsic germanium (i-Ge) detector These detectors require use of the highest purity germanium, having electrically active impurities of density no

20 Electromagnetic Properties and Nuclear Decay

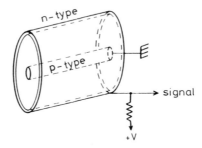

Fig. 2.12 Cylindrical Ge(Li) detector.

greater than approximately 2×10^{10} cm^{-3}. If used at temperatures of order 77 K, the temperature of liquid nitrogen, thermal carrier densities are negligible, and the material is an excellent insulator. If a voltage is applied across the detector, the carriers created by γ-ray absorption can be detected as a current pulse through a load resistor. The integrated charge pulse is proportional to the number of carriers created in this event and is proportional to the energy of the absorbed γ-ray.

Ge(Li) detector These detectors are p-i-n diodes in which the bulk of the material is insulating and are made by allowing lithium to drift in under controlled conditions such that the lithium donors exactly compensate the acceptors in the originally p-type material. Extremely large volumes (up to approximately 150 cm^3) are obtainable in cylindrical geometry (Fig. 2.12). These devices are made by diffusing in lithium from the outer rim of a cylinder of originally p-type germanium under an applied voltage. The diffusion continues until the entire cylinder is compensated apart from a small core at the center. A lithium excess remains on the outer rim making it n-type. The p-type core and n-type rim serve as the electrical contacts of the p-i-n structure.

Ge detectors are distinguished by excellent energy resolution ($\Delta E \approx$ 2 keV for 1 MeV γ-rays, see Fig. 2.11). The time resolution is typically 5 ns.

3
Hyperfine Interactions

Up to now we have considered only the electromagnetic properties of free nuclei and emission of γ-radiation. We now consider the central problem of the interaction between a nucleus and an electric or magnetic field. Such fields are produced in solids by electrons and other nuclei in the vicinity of the nucleus. In addition, external fields, for example an externally applied magnetic field, may be present. The interaction of a nucleus with these fields is called a hyperfine interaction. This interaction was discovered in atomic spectroscopy and its effects on atomic spectra have been extensively investigated. However, the hyperfine field also affects the nucleus and makes determination of internal fields possible in solids through measurement of nuclear properties. Detection of hyperfine interactions is the goal of most of the experimental methods discussed in this book.

3.1 Magnetic Interaction

The magnetic nuclear dipole moment $\boldsymbol{\mu}$ interacts with the magnetic flux density $\boldsymbol{B}$ at the position of the nucleus. The interaction energy is

$$E_{\text{magn}} = -\boldsymbol{\mu} \cdot \boldsymbol{B} \tag{3.1}$$

This extra energy lifts the degeneracy of the nuclear M states and induces a precession of the nuclear spin. Both aspects will be considered in the following.

3.1.1 Level Splitting

In classical physics, the magnetic interaction energy given in Eq. (3.1) depends on the angle between $\boldsymbol{\mu}$ and $\boldsymbol{B}$ and therefore should be continuously distributed over the range between its maximum and minimum energy. In quantum physics, quantization of the angular momentum allows only certain

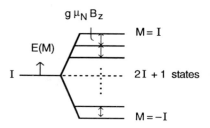

Fig. 3.1 Nuclear level splitting in a **B** field (nuclear Zeeman effect).

directional possibilities for **μ** with respect to **B**. Choosing the z axis parallel to **B**, one obtains

$$E_{\text{magn}} = \langle I, M | -\mu_z B_z | I, M \rangle$$
$$= -\gamma B_z \langle I, M | I_z | I, M \rangle = -\gamma B_z \hbar M \tag{3.2}$$

The energy difference between adjacent M levels is

$$E_{\text{magn}}(M+1) - E_{\text{magn}}(M) = -\gamma \hbar B_z = -g\mu_N B_z \tag{3.3}$$

This means that the splitting is the same between all neighboring M levels (see Fig. 3.1).

3.1.2 Nuclear Spin Precession

The magnetic interaction not only induces a level splitting but also causes a time dependence in the expectation values of nuclear properties. The time dependent wave function is

$$\psi(t) = \Lambda(t)\psi(0) \tag{3.4}$$

where $\Lambda(t) = \exp(-i\mathcal{H}t/\hbar)$. $\mathcal{H} = -\gamma I_z B_z$ is the Hamiltonian operator and $\Lambda(t)$ is the time-evolution operator. Substituting these expressions in Eq. (3.4) one obtains

$$\Lambda(t) = \exp[-i(-\gamma I_z B_z)t/\hbar] = \exp[-i(-\gamma B_z t)I_z/\hbar] \tag{3.5}$$

$\Lambda(t)$ in Eq. (3.5) has the form of a rotation operator around the z axis and can be written

$$\Lambda(t) = \exp(-i\alpha I_z/\hbar) \tag{3.6}$$

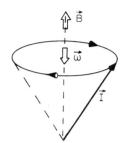

Fig. 3.2 Precession of angular momentum I in magnetic field B. In quantum mechanics I must be replaced by the expectation value $\langle I \rangle$. The rotation direction shown in the figure is for a positive g-factor.

where α is the rotation angle. From Eqs. (3.6) and (3.4) one can see that the classical Larmor frequency

$$\omega_L = -\gamma B_z = -g(\mu_N/\hbar) B_z \tag{3.7}$$

appears in the time development of the state. The expectation value of the angular momentum I calculated with wave functions of the form in Eq. (3.4) precesses around B with frequency ω_L (Fig. 3.2). This can be demonstrated by a simple example.

EXAMPLE Nuclear spin precession for $I = 1/2$ ($L = 0$, $S = 1/2$)

The expectation value $\langle I \rangle = \langle S \rangle$ must be calculated as a function of time. We choose the coordinate system so the z axis is parallel to B and consider the special case where the initial polarization is parallel to the x axis. For the spin vector, we use the following representation

$$S_x = \frac{1}{2}(S_+ + S_-)$$

$$S_y = \frac{1}{2i}(S_+ - S_-) \tag{3.8}$$

$$S_z = S_z$$

If we represent the spin-up wave functions by $|+\rangle$ and spin-down by $|-\rangle$, the following equations hold

$$\begin{array}{lll} S_+|+\rangle = 0 & S_-|+\rangle = \hbar|-\rangle & S_z|+\rangle = (\hbar/2)|+\rangle \\ S_+|-\rangle = \hbar|+\rangle & S_-|-\rangle = 0 & S_z|-\rangle = (-\hbar/2)|-\rangle \end{array} \tag{3.9}$$

We show first that the initial state

$$\psi(t=0) = \frac{1}{\sqrt{2}}(|+\rangle + |-\rangle) \tag{3.10}$$

has spin polarization in the x direction:

$$\langle \psi(t=0)|S_x|\psi(t=0)\rangle = \frac{1}{2}(\langle-|+\langle+|)\frac{1}{2}(S_+ + S_-)(|+\rangle + |-\rangle)$$

$$= \frac{1}{4}(\hbar + \hbar) = \frac{\hbar}{2} \qquad (3.11\text{a})$$

$$\langle \psi(t=0)|S_y|\psi(t=0)\rangle = \frac{1}{2}(\langle-|+\langle+|)\frac{1}{2i}(S_+ - S_-)(|+\rangle + |-\rangle)$$

$$= \frac{1}{4i}(\hbar - \hbar) = 0 \qquad (3.11\text{b})$$

$$\langle \psi(t=0)|S_z|\psi(t=0)\rangle = \frac{1}{2}(\langle-|+\langle+|)S_z(|+\rangle + |-\rangle)$$

$$= \frac{1}{2}\left(\frac{\hbar}{2} - \frac{\hbar}{2}\right) = 0 \qquad (3.11\text{c})$$

For $t=0$ we obtain

$$\langle S_x \rangle = \frac{\hbar}{2} \qquad \langle S_y \rangle = 0 \qquad \langle S_z \rangle = 0 \qquad (3.12)$$

i.e. the spin is polarized in the x direction.

We now consider nonzero times. Using the time-development operator of Eq. (3.5), we obtain for the wave function

$$\psi(t) = \exp(-i\omega_L t S_z/\hbar)\psi(t=0) \qquad (3.13)$$

where the Hamiltonian operator (here $I = S$, and therefore $g = g_S$)

$$\mathcal{H} = -\mu_z B_z = -\gamma B_z S_z = \omega_L S_z \qquad (3.14)$$

was used. Here, ω_L is the Larmor frequency defined in Eq. (3.7). Using Eqs. (3.10) and (3.13) we obtain

$$\psi(t) = \frac{1}{\sqrt{2}}\left[\exp\left(\frac{-i\omega_L t}{2}\right)|+\rangle + \exp\left(\frac{+\omega_L t}{2}\right)|-\rangle\right] \qquad (3.15)$$

The expectation values of the components of $\langle S \rangle$ then become

$$\langle \psi(t)|S_x|\psi(t)\rangle = \langle \psi(t)|\frac{1}{2}(S_+ + S_-)|\psi(t)\rangle$$

$$= \frac{1}{4}\left[\langle+|\exp\left(\frac{+i\omega_L t}{2}\right) + \langle-|\exp\left(\frac{-i\omega_L t}{2}\right)\right](S_+ + S_-)$$

$$\times \left[\exp\left(\frac{-i\omega_L t}{2}\right)|+\rangle + \exp\left(\frac{+i\omega_L t}{2}\right)|-\rangle \right]$$

$$= \frac{\hbar}{4}[\exp(i\omega_L t) + \exp(-i\omega_L t)]$$

or

$$\langle \psi(t)|S_x|\psi(t)\rangle = \frac{\hbar}{2}\cos(\omega_L t) \quad (3.16a)$$

$$\langle \psi(t)|S_y|\psi(t)\rangle = \frac{\hbar}{2}\sin(\omega_L t) \quad (3.16b)$$

$$\langle \psi(t)|S_z|\psi(t)\rangle = 0 \quad (3.16c)$$

The calculations leading to (3.16b) and (3.16c) are not shown but are analogous to those shown for (3.16a). Thus we find that the spin precesses in the xy plane with frequency ω_L.

3.2 Electric Interaction

In a solid, a nucleus is surrounded by electrical charges which produce a potential $\Phi(r)$. Classically, the interaction energy of a nuclear charge distribution $\rho(r)$ in an external potential $\Phi(r)$ is given by

$$E_{\text{electr}} = \int \rho(r)\Phi(r)\,d^3r \quad (3.17)$$

where

$$\int \rho(r)\,d^3r = Ze \quad (3.18)$$

is the nuclear charge. It is convenient to expand the electrical potential in a Taylor series around $r = 0$:

$$\Phi(r) = \Phi_0 + \sum_{\alpha=1}^{3}\left(\frac{\partial \Phi}{\partial x_\alpha}\right)_0 x_\alpha + \frac{1}{2}\sum_{\alpha,\beta}\left(\frac{\partial^2 \Phi}{\partial x_\alpha \partial x_\beta}\right)_0 x_\alpha x_\beta + \ldots \quad (3.19)$$

where x_α and x_β are Cartesian coordinates. With this expansion, one obtains for the energy

$$E_{\text{electr}} = E^{(0)} + E^{(1)} + E^{(2)} + \ldots$$

where

$$E^{(0)} = \Phi_0 \int \rho(r) \, d^3r$$

$$E^{(1)} = \sum_{\alpha=1}^{3} \left(\frac{\partial \Phi}{\partial x_\alpha}\right)_0 \int \rho(r) x_\alpha \, d^3r \qquad (3.20)$$

$$E^{(2)} = \frac{1}{2}\sum_{\alpha,\beta} \left(\frac{\partial^2 \Phi}{\partial x_\alpha \partial x_\beta}\right)_0 \int \rho(r) x_\alpha x_\beta \, d^3r$$

The meaning of each of the three terms will be examined individually. Using Eq. (3.18), the first term can be rewritten $E^{(0)} = \Phi_0 Z e$. Since Φ_0 is the potential at the origin ($r = 0$), $E^{(0)}$ is the Coulomb energy of a point-like charge distribution in the external potential. This term is the same for all isotopes of a given element. $E^{(0)}$ contributes to the potential energy of the crystal lattice but is irrelevant to the present discussions.

The second term $E^{(1)}$ represents an electric dipole interaction between the electric field $\mathbf{E} = -\nabla \Phi$ at the origin and the electric dipole moment of the nuclear charge distribution. The quantum mechanical expectation value of the nuclear electric dipole moment is zero, since the nuclear states have definite parity. Thus $E^{(1)}$ is zero.

Therefore we are left with the third term $E^{(2)}$. The quantities

$$\left(\frac{\partial^2 \Phi}{\partial x_\alpha \partial x_\beta}\right)_0 =: \Phi_{\alpha\beta} \qquad (3.21)$$

form a symmetric 3×3 matrix which can be diagonalized by a coordinate rotation. After the diagonalization one obtains

$$E^{(2)} = \frac{1}{2}\sum_\alpha \Phi_{\alpha\alpha} \int \rho(r) x_\alpha^2 \, d^3r$$

$$= \frac{1}{6}\sum_\alpha \Phi_{\alpha\alpha} \int \rho(r) r^2 \, d^3r + \frac{1}{2}\sum_\alpha \Phi_{\alpha\alpha} \int \rho(r)\left(x_\alpha^2 - \frac{r^2}{3}\right) d^3r \qquad (3.22)$$

where $r^2 = x_1^2 + x_2^2 + x_3^2$ was used. The separation into the two separate terms in the second line is explained below.

The electrostatic potential $\Phi(r)$ obeys the Poisson equation. At the nucleus we have

$$(\Delta\Phi)_0 = \sum_\alpha \Phi_{\alpha\alpha} = \frac{e}{\varepsilon_0}|\psi(0)|^2 \qquad (3.23)$$

where $-e|\psi(0)|^2$ is the charge density and $|\psi(0)|^2$ is the probability density of electrons (s electrons) at the nucleus. Therefore we obtain for $E^{(2)}$

$$E^{(2)} = E_C + E_Q$$

where

$$\begin{aligned} E_C &= \frac{e}{6\varepsilon_0}|\psi(0)|^2 \int \rho(r) r^2 \, d^3r \\ E_Q &= \frac{1}{2}\sum_\alpha \Phi_{\alpha\alpha} \int \rho(r)\left(x_\alpha^2 - \frac{r^2}{3}\right) d^3r \end{aligned} \qquad (3.24)$$

3.2.1 The Monopole Term E_C

From Eq. (3.24) one sees that the monopole term E_C depends only on the mean square nuclear radius

$$\langle r^2 \rangle := \frac{1}{Ze} \int \rho(r) r^2 \, d^3r \qquad (3.25)$$

E_C gives a shift but no splitting of the levels. Thus this monopole term describes the electrostatic interaction of an *extended* nucleus with the electrons at the nuclear site. This effect is illustrated in Fig. 3.3.

The monopole term can be written

$$E_C = \frac{Ze^2}{6\varepsilon_0}|\psi(0)|^2 \langle r^2 \rangle \qquad (3.26)$$

In atomic physics, this term is responsible for the isotope shift. It causes spectral lines of two isotopes with different nuclear radii to be slightly different. This monopole term is also responsible for the Mössbauer isomer shift discussed in Section 4.4.

3.2.2 The Electric Quadrupole Interaction E_Q

The expression for E_Q in Eq. (3.24) contains the integral $\int \rho(r)(x_\alpha^2 - r^2/3) \, d^3r$. For $x_\alpha = z$, this integral is $(e/3)$ times the classical quadrupole moment

28   Hyperfine Interactions

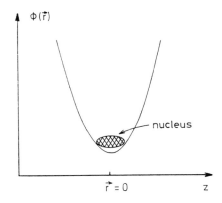

Fig. 3.3 Graphical explanation of the monopole and quadrupole terms. *Monopole term*: The nucleus is located in a potential minimum. The outer portion of the nucleus is consequently at a somewhat higher potential than the central part (i.e. the potential a point charge nucleus would experience). The energy increase compared to that of a point charge depends on the curvature of the potential and the extent of the nuclear charge distribution. *Quadrupole term*: If the curvatures of the potential in the three spatial directions are not equal, then the interaction energy depends on the orientation of the deformed nucleus. The smallest energy is obtained if the longest nuclear axis is parallel to the direction of weakest curvature.

of the nucleus (see Eq. (2.13)). We can now write

$$E_Q = \frac{e}{6}\sum_\alpha \Phi_{\alpha\alpha} Q_{\alpha\alpha}$$

where (3.27)

$$Q_{\alpha\alpha} = \frac{1}{e}\int \rho(r)(3x_\alpha^2 - r^2)\,\mathrm{d}^3 r$$

In the following, we discuss the terms $\Phi_{\alpha\alpha}$ which contribute to the trace ($\Delta\Phi = \sum \Phi_{\alpha\alpha}$). We define

$$\Phi_{\alpha\alpha} = V_{\alpha\alpha} + \frac{1}{3}(\Delta\Phi) \tag{3.28}$$

The matrix $V_{\alpha\alpha}$ defined by this equation is traceless. If we substitute (3.28) in Eq. (3.27), the portion $(1/3)(\Delta\Phi)$ does not contribute to E_Q since $\sum Q_{\alpha\alpha}$ is zero. Thus we have (see also Fig. 3.3)

$$E_Q = \frac{e}{6}\sum_\alpha V_{\alpha\alpha} Q_{\alpha\alpha} \tag{3.29}$$

Electric Interaction

$V_{\alpha\alpha}$ is called the electric field gradient; only charges not at the nuclear site contribute to $V_{\alpha\alpha}$. For spherically symmetric charge distributions (s electrons), $V_{xx} = V_{yy} = V_{zz}$. Since $\sum V_{\alpha\alpha} = 0$, all components of the electric field gradient are zero and cannot contribute to the energy E_Q.

Because $\sum V_{\alpha\alpha} = 0$, the electric field gradient is completely described by two parameters. By appropriate choice of the principal axis system, one can have $|V_{zz}| \geq |V_{yy}| \geq |V_{xx}|$. Generally one chooses to describe the electric field gradient by V_{zz} and the asymmetry parameter η

$$\eta = \frac{V_{xx} - V_{yy}}{V_{zz}} \qquad (3.30)$$

The quantity E_Q is a product of the nuclear electric quadrupole moment and the electric field gradient at the nuclear position (nuclear quadrupole interaction). To calculate E_Q we will represent the nuclear quadrupole moment and the electric field gradient by spherical tensors. Tensor algebra allows a particularly elegant treatment of E_Q. In Appendix A.2 some important tensor properties are summarized.

The nuclear electric quadrupole tensor is defined as (compare with Eq. (2.14))

$$Q_{2q} = Zr^2 Y_2^q \qquad (3.31)$$

For the electric field gradient the tensor components are

$$V_{20} = \frac{1}{4}\sqrt{\frac{5}{\pi}} V_{zz}$$

$$V_{2\pm1} = \mp\frac{1}{2}\sqrt{\frac{5}{6\pi}}(V_{xz} \pm V_{yz}) \qquad (3.32)$$

$$V_{2\pm2} = \frac{1}{4}\sqrt{\frac{5}{6\pi}}(V_{xx} - V_{yy} \pm 2iV_{xy})$$

In the principal axis system,

$$V_{20} = \frac{1}{4}\sqrt{\frac{5}{\pi}} V_{zz} \qquad V_{2\pm1} = 0$$

$$V_{2\pm2} = \frac{1}{4}\sqrt{\frac{5}{6\pi}}(V_{xx} - V_{yy}) = \frac{1}{4}\sqrt{\frac{5}{6\pi}}\eta V_{zz} \qquad (3.33)$$

In the tensor representation, the electric quadrupole interaction is given by (Matthias *et al.*, 1962)

$$E_Q = \frac{4\pi}{5}\sum_q (-)^q eQ_{2q} V_{2-q} \tag{3.34}$$

Limiting ourselves to axially symmetric electric field gradients ($V_{xx} = V_{yy}$, or $\eta = 0$) we obtain

$$E_Q = \sqrt{\frac{\pi}{5}} eQ_{20} V_{zz} \tag{3.35}$$

In the quantum formulation, Q_{20} is an operator acting on the nuclear wave function. The calculation of E_Q requires that the matrix elements of Q_{20} be known,

$$E_Q = \sqrt{\frac{\pi}{5}} V_{zz} e \langle I, M | Q_{20} | I, M \rangle \tag{3.36}$$

Applying the Wigner–Eckart theorem gives (see Appendix A.3)

$$\langle I, M | Q_{20} | I, M \rangle = (-)^{I-M} \begin{pmatrix} I & 2 & I \\ -M & 0 & M \end{pmatrix} \langle I \| Q_2 \| I \rangle \tag{3.37}$$

By definition

$$Q = 4\sqrt{\frac{\pi}{5}} \langle I, I | Q_{20} | I, I \rangle = 4\sqrt{\frac{\pi}{5}} \begin{pmatrix} I & 2 & I \\ -I & 0 & I \end{pmatrix} \langle I \| Q_2 \| I \rangle \tag{3.38}$$

Combining these two equations,

$$E_Q = \frac{1}{4} V_{zz} (-)^{I-M} \frac{\begin{pmatrix} I & 2 & I \\ -M & 0 & M \end{pmatrix}}{\begin{pmatrix} I & 2 & I \\ -I & 0 & I \end{pmatrix}} eQ \tag{3.39}$$

$$E_Q = \frac{3M^2 - I(I+1)}{4I(2I-1)} eQV_{zz}$$

The transition energy between two sublevels M and M' is

$$E_Q(M) - E_Q(M') = \frac{3eQV_{zz}}{4I(2I-1)}|M^2 - M'^2| = 3|M^2 - M'^2|\hbar\omega_Q \quad (3.40)$$

where we have introduced the quadrupole frequency

$$\omega_Q = \frac{eQV_{zz}}{4I(2I-1)\hbar} \quad (3.41)$$

Because $(M^2 - M'^2) = (M + M')(M - M')$, the quantity $|M^2 - M'^2|$ is always an integer. Thus all transition frequencies are integer multiples of the lowest transition frequency,

$$\begin{aligned}\omega_Q^0 &= 6\omega_Q \quad \text{(for half-integer nuclear spin)} \\ \omega_Q^0 &= 3\omega_Q \quad \text{(for integer nuclear spin)}\end{aligned} \quad (3.42)$$

Generally one uses the derived quantity ν_Q

$$\nu_Q = eQV_{zz}/h \quad (3.43)$$

EXAMPLE Electric quadrupole splitting for nuclear spin 5/2
Using Eq. (3.39) the energy for different M states can be calculated. Because of the M^2 dependence of E_Q, $+M$ and $-M$ states are degenerate. One obtains

$$\begin{aligned} E_Q(M = \pm 1/2) &= -\frac{1}{5}eQV_{zz} \\ E_Q(M = \pm 3/2) &= -\frac{1}{20}eQV_{zz} \\ E_Q(M = \pm 5/2) &= +\frac{1}{4}eQV_{zz} \end{aligned} \quad (3.44)$$

Fig. 3.4 shows the quadrupole splitting of an $I = 5/2$ state. The transition frequencies ω_1, ω_2, and $\omega_3 = \omega_1 + \omega_2$ are given by

$$\begin{aligned} \omega_1 &:= \omega_Q^0 = \frac{E_Q(\pm 3/2) - E_Q(\pm 1/2)}{\hbar} = \frac{3eQV_{zz}}{20\hbar} = 6\omega_Q \\ \omega_2 &= \frac{E_Q(\pm 5/2) - E_Q(\pm 3/2)}{\hbar} = \frac{6eQV_{zz}}{20\hbar} = 12\omega_Q \\ \omega_3 &= \omega_1 + \omega_2 = \frac{9eQV_{zz}}{20\hbar} = 18\omega_Q \end{aligned} \quad (3.45)$$

32   Hyperfine Interactions

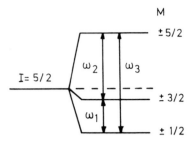

Fig. 3.4 Energy splitting of an $I = 5/2$ nuclear level under the influence of an axially symmetric quadrupole interaction.

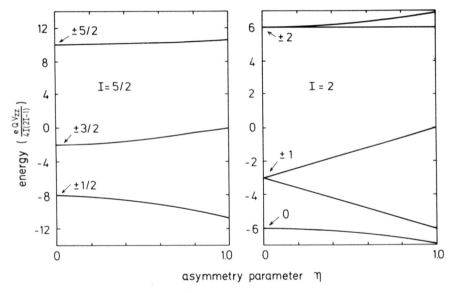

Fig. 3.5 Energy splitting of an $I = 5/2$ and $I = 2$ nuclear state as a function of the asymmetry parameter η (V_{zz} = constant). The M values shown in the figure are strictly valid only for $\eta = 0$.

The transition frequencies $\omega_1 : \omega_2 : \omega_3$ are in the ratio $1 : 2 : 3$.

Thus far we have limited our discussion to the axially symmetric case. For $\eta \neq 0$, the matrix elements for E_Q cannot be calculated analytically in general. The Hamiltonian must be diagonalized numerically. In Fig. 3.5 the energy levels for nuclei of spin $I = 5/2$ and $I = 2$ are shown as a function of asymmetry parameter η for a fixed V_{zz}.

4
Mössbauer Effect

4.1 Principles

A γ-ray can be absorbed by an atom only if its energy is equal to an excitation energy of the nucleus. One might think that this will always occur if the emission and absorption are by the same nuclear transition. That is not the case, however, because the emitting nucleus recoils and the γ-ray energy is smaller than the nuclear transition energy. A similar effect occurs on absorption, because momentum conservation requires that part of the γ-ray energy be converted to kinetic energy of the absorbing nucleus. In 1957, R. Mössbauer (Mössbauer, 1958) found that the energy reductions are avoided if the emitting and absorbing nuclei are part of a crystalline solid. For many such cases, the recoil momentum during emission and absorption is transferred to the crystal as a whole. Because of the large mass of the crystal, negligible energy is lost due to recoil. This process is called recoilless emission and absorption of γ-radiation and is referred to as the Mössbauer effect.

The principle of the Mössbauer effect is illustrated in Fig. 4.1. An excited nucleus decays to the ground state by emitting a γ-ray. The emitted γ-ray can be absorbed by another nucleus of the same kind if both processes are recoilless. Since the absorption in general can occur only for a nucleus in the

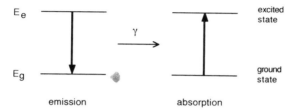

Fig. 4.1 Recoilless emission and absorption of a γ-ray (Mössbauer effect).

ground state, only a γ-ray emitted in a transition to the ground state is useful for practical application of the Mössbauer effect.

For recoilless emission and absorption, the energy uncertainty is limited only by the natural linewidth of the nuclear decay. This uncertainty is many orders of magnitude smaller than the recoil energy or Doppler shift of free atoms. In the remainder of this section, the order of magnitude of these effects is discussed in more detail.

Natural linewidth

A nuclear level with mean lifetime τ_N (or halflife $t_{1/2} = \tau_N \ln 2$) has energy uncertainty given by $\Gamma = \hbar/\tau_N$. Γ represents the natural linewidth.

It can be shown that the frequency spectrum of the emitted γ-rays has a Lorentz distribution centered at ω_0 and halfwidth $\Gamma/\hbar$ (Fig. 4.2),

$$I(\omega) = \frac{I_0}{1 + [(\omega - \omega_0)2\hbar/\Gamma]^2} \tag{4.1}$$

where $I(\omega)$ is the intensity of the radiation at frequency ω.

For the frequently utilized nucleus, ^{57}Fe, the energy of the transition $\hbar\omega_0$ is 14.4 keV, and the lifetime τ_N is 1.41×10^{-7} s, so the linewidth is

$$\Gamma = \hbar/\tau_N = 4.7 \times 10^{-9} \text{ eV} \tag{4.2}$$

and the relative energy uncertainty

$$\Gamma/\hbar\omega_0 = 3.3 \times 10^{-13} \tag{4.3}$$

By observation of resonant absorption, it is possible to measure energies

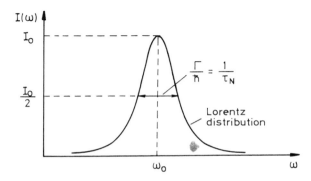

Fig. 4.2 Intensity distribution of the emitted γ-radiation from a nuclear decay.

Recoil and Doppler shift

Emission of a γ-ray leads to a recoil of the nucleus and reduction of the γ-ray energy. The consequences of this effect on the frequency spectrum of the emitted γ-rays is discussed for a monatomic gas in thermal equilibrium. The energy of the excited nucleus before the emission of the γ-ray is

$$E_{\text{before}} = E_e + \frac{p^2}{2M} \tag{4.4}$$

where $p = Mv$ is the momentum and M the mass of the atom. After the emission, the momentum is reduced by $\hbar k$ (i.e. by the momentum of the emitted γ-ray), and the internal energy of the nucleus becomes E_g. Thus

$$E_{\text{after}} = E_g + \frac{(p - \hbar k)^2}{2M} \tag{4.5}$$

and the energy difference is

$$E_{\text{before}} - E_{\text{after}} := \hbar\omega = \hbar\omega_0 + \hbar(\mathbf{k} \cdot \mathbf{v}) - \frac{\hbar^2 k^2}{2M} \tag{4.6}$$

The term $\hbar(\mathbf{k} \cdot \mathbf{v})$ is the (velocity-dependent) Doppler effect. For ^{57}Fe at room temperature, the Doppler shift is of the order of 10^{-2} eV.

The constant recoil energy term $\hbar^2 k^2 / 2M$ is 2×10^{-3} eV for ^{57}Fe. These energies are approximately six orders of magnitude larger than the natural linewidth of 4.7×10^{-9} eV. The frequency spectrum emitted by a monatomic gas is shown schematically in Fig. 4.3.

In an absorber, the situation is reversed: The center of the absorption spectrum is shifted to frequencies above ω_0. The absorption curve is a mirror image around ω_0 of the spectrum shown in Fig. 4.3.

Only a small resonant absorption would occur if both emitter and absorber are gaseous, because only the wings of the emission and absorption lines overlap. In addition, this process is not very selective since the absorption takes place over a wide frequency range. The main point of the Mössbauer effect is that a large part of the emission and absorption spectrum is condensed into a small region around ω_0.

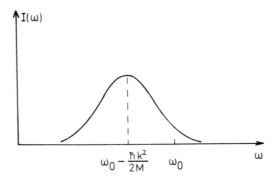

Fig. 4.3 Emission spectrum of a monatomic gas in thermal equilibrium. The emission spectrum is shifted from ω_0 by the recoil energy and is broadened due to the Maxwell velocity distribution of the gas atoms.

4.2 The Debye–Waller Factor

If the emitting atom is in a solid, it is relatively strongly bound; the typical energy required for displacement of the atom to another position is 20 eV. Since the recoil energy following emission of a γ-ray is not sufficient to displace the atom, it can do nothing more than set the atom into vibrational motion. If the vibrational motion is not altered by the emission, then the recoil energy must be absorbed by the solid as a whole. Since the mass of the crystal is approximately 10^{20} times larger than that of a single atom, the recoil energy is negligible. In that case, one obtains an unbroadened, unshifted emission (Mössbauer) line.

The ratio of the unshifted γ-ray emission to the total is called the Debye–Waller factor f. The Debye–Waller factor is temperature dependent and is largest at $T = 0$. However, even at absolute zero, $f < 1$ because of zero-point energy.

A deeper understanding of the Mössbauer effect requires a more thorough understanding of the Debye–Waller factor. The derivation of the Debye–Waller factor demonstrates why the Mössbauer fraction is large. To show this essential point as clearly as possible, we present here an easily understandable semiclassical explanation of the Debye–Waller factor and, consequently, of the Mössbauer effect. For a quantum mechanical derivation we refer the reader to Wegener (1966).

Semiclassical derivation of the Debye–Waller factor

As a simple classical model, we consider a nucleus continuously emitting an electromagnetic wave with frequency $\omega_0 = E_\gamma/\hbar$ and simultaneously vibrating around an equilibrium position. Of course in reality the γ-rays are emitted in quanta. Classically there is no recoil, since the radiation pressure on a dipole is zero by symmetry.

Initially, we will limit consideration to a single vibration frequency Ω. This corresponds to the Einstein model of vibrations in solids. One obtains for the electric field $E(t)$ of the emitted radiation,

$$E(t) = E_0 \exp\{-i[\omega_0 t + kx(t)]\} \tag{4.7}$$

where the atomic motion is assumed to be $x(t) = a \sin \Omega t$; a is the amplitude of the oscillation. With the substitution of $x(t)$ into Eq. (4.7) and expansion of the exponential function, we obtain

$\exp(-i\omega_0 t) \exp[-i kx(t)]$

$$= \exp(-i\omega_0 t) \exp(-i\,ka \sin \Omega t)$$
$$= \exp(-i\omega_0 t)\left(1 - i\,ka \sin \Omega t - \frac{k^2 a^2}{2} \sin^2 \Omega t + \ldots\right) \tag{4.8}$$

Using the relation

$$\sin^n \Omega t = \frac{1}{(2i)^n}[\exp(i\Omega t) - \exp(-i\Omega t)]^n \tag{4.9}$$

yields

$$\exp(-i\omega_0 t) \exp[-i\,kx(t)] = \left(1 - \frac{k^2 a^2}{4} + \ldots\right)\exp(-i\omega_0 t)$$
$$+ \left(-\frac{ka}{2} + \ldots\right)\{\exp[-i(\omega_0 - \Omega)t] - \exp[-i(\omega_0 + \Omega)t]\}$$
$$+ \left(\frac{k^2 a^2}{8} + \ldots\right)\{\exp[-i(\omega_0 - 2\Omega)t] + \exp[-i(\omega_0 + 2\Omega)t]\}$$
$$+ \ldots \tag{4.10}$$

The effect of the atomic motion is to reduce the intensity of the central line by the factor $(1 - k^2 a^2/4 + \ldots)$ and to produce sidebands with frequencies $\omega_0 \pm \Omega$, $\omega_0 \pm 2\Omega$, ... The Debye–Waller factor is the unshifted line intensity, which is reduced due to the atomic motion. The frequency spectrum is shown schematically in Fig. 4.4.

38 Mössbauer Effect

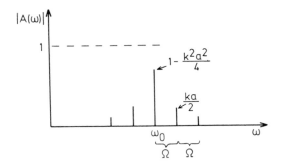

Fig. 4.4 Central line and sideband frequencies for a classical model illustrating the Debye–Waller factor and the Mössbauer effect.

Summing the contributions to the intensity of the unshifted line given in Eq. (4.8), one obtains

$$A(\omega_0) = 1 - \frac{k^2 a^2}{4} + \ldots = \mathcal{J}_0(ka) \tag{4.11}$$

where $\mathcal{J}_0(ka)$ is the Bessel function of zero order. The Debye–Waller factor is the absolute square of $A(\omega_0)$, i.e. $\mathcal{J}_0^2(ka)$.

Note that the terms in the middle of Eq. (4.11) are the first expansion terms of the exponential function $\exp(-k^2 a^2/4)$. However, $k^2 a^2/4$ (a being the toal oscillation amplitude of the atom) is, in general, not small compared to 1 and therefore $A(\omega_0)$ cannot be replaced by this exponential function. The fact that nevertheless the exponential and not the Bessel function describes the Debye–Waller factor will be explained in the following (see Wegener, 1966).

In a real crystal there are many different normal modes (phonons) and the Debye–Waller factor is described by the product $\prod \mathcal{J}_0^2(ka_n)$ where the product has to be taken over all normal modes. The amplitudes a_n of the individual normal modes are very small and therefore the condition $k^2 a_n^2/4 \ll 1$ is very well satisfied. Thus we have

$$f = \prod \mathcal{J}_0^2(ka_n) \approx \prod \left(1 - \frac{k^2 a_n^2}{4}\right)^2 \approx \prod \left(1 - \frac{k^2 a_n^2}{2}\right)$$

$$\approx \prod \exp\left(-\frac{k^2 a_n^2}{2}\right) = \exp\left(-\frac{k^2 \sum a_n^2}{2}\right) = \exp(-k^2 \langle x^2 \rangle) \tag{4.12}$$

The relationship $\langle x^2 \rangle = \sum a_n^2/2$ relating the average square displacement to the oscillation amplitudes of one-dimensional oscillators is used above.

For the isotropic three-dimensional oscillator, the average square displacement $\langle u^2 \rangle = \langle x^2 \rangle + \langle y^2 \rangle + \langle z^2 \rangle = 3\langle x^2 \rangle$. One obtains

$$f = \exp\left(-\frac{k^2 \langle u^2 \rangle}{3}\right) \quad (4.13)$$

This equation is valid for the general case where the atomic motion is a superposition of many frequencies (phonons); each frequency produces its own sidebands, which form a quasi-continuous distribution. The important point is that each phonon contributes to the central line whose intensity therefore dominates the spectrum.

The classically derived Debye–Waller factor given by Eq. (4.13) is also obtained by a rigorous quantum mechanical derivation if $\langle u^2 \rangle$ is taken as the quantum mechanical expectation value.

This Debye–Waller factor is well known from diffraction experiments. The intensities of X-ray, neutron, and electron diffraction lines are given by

$$I_{hkl} = I^0_{hkl} \exp\left(-\frac{g^2 \langle u^2 \rangle}{3}\right) \quad (4.14)$$

where g is the scattering vector which corresponds to a reciprocal lattice vector G_{hkl} at a point of constructive interference (Bragg condition).

The differences between the above classical model and the rigorous quantum mechanical treatment are illustrated graphically in Fig. 4.5. The classical model yields a symmetric distribution of the quasi-continuous background around the central line. Quantum mechanically, one finds that

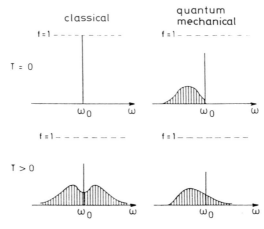

Fig. 4.5 Comparison of the frequency spectrum derived from classical and quantum mechanical treatments. Figure adapted from Wegener (1966).

the mean of the background is always shifted to lower values relative to the unshifted line. The shift corresponds to the recoil energy of a free emitting atom.

At $T = 0$, one obtains classically the value $f = 1$ since no oscillations are excited. Quantum mechanically $f < 1$, since $\langle u^2 \rangle \neq 0$ because of zero-point motion. However, at $T = 0$ there are no emission lines above ω_0 since there are no phonons to be absorbed by the emitted γ-ray.

In the following we will calculate $\langle x^2 \rangle$ for the Einstein model where all atoms vibrate with the single frequency Ω. The total energy of a harmonic oscillator with n excited quanta is

$$E_n = \hbar\Omega(n + \tfrac{1}{2}) \tag{4.15}$$

and the probability that n quanta are excited is given by

$$P_n = \frac{\exp(-n\hbar\Omega/k_B T)}{\sum_{n=0}^{\infty} \exp(-n\hbar\Omega/k_B T)} \tag{4.16}$$

where k_B is the Boltzmann constant. The time-average square displacement $\overline{x_n^2}$ of an oscillator in the quantum state n is obtained by equating the average potential energy $M\Omega^2 x_n^2/2$ with half the total energy

$$\frac{1}{2}M\Omega^2\overline{x_n^2} = \frac{1}{2}\hbar\Omega(n + \tfrac{1}{2}) \tag{4.17}$$

Using the occupation probability given by Eq. (4.16) and averaging over all states n, we obtain

$$\langle \overline{x^2} \rangle_E = \frac{\hbar}{M\Omega}\left(\frac{1}{2} + \sum_{n=0}^{\infty} nP_n\right) \tag{4.18}$$

The index E is used to indicate that this formula applies to the Einstein model.

The expression $\sum nP_n$ yields the Bose–Einstein distribution function. One then obtains, with the bar indicating time averaging omitted for clarity,

$$\langle x^2 \rangle_E = \frac{\hbar}{2M\Omega}\left[1 + \frac{2}{\exp(\hbar\Omega/k_B T) - 1}\right] \tag{4.19}$$

Up to this point only one phonon frequency has been considered. We now generalize to a phonon distribution function $Z(\Omega)$ having normalization $\int Z(\Omega)\,d\Omega = 3N$, where N is the number of molecules in the solid. Thus

$$\langle x^2 \rangle = \frac{1}{3N} \int_0^\infty Z(\Omega) \langle x^2 \rangle_E \, d\Omega$$

$$= \frac{\hbar}{6MN} \int_0^\infty \frac{Z(\Omega)}{\Omega} \left[1 + \frac{2}{\exp(\hbar\Omega/k_B T) - 1} \right] d\Omega \quad (4.20)$$

Inserting $\langle x^2 \rangle = \langle u^2 \rangle / 3$ in Eq. (4.13) yields for the Debye–Waller factor

$$f(T) = \exp \left\{ -\frac{\hbar k^2}{6MN} \int_0^\infty \frac{Z(\Omega)}{\Omega} \left[1 + \frac{2}{\exp(\hbar\Omega/k_B T) - 1} \right] d\Omega \right\} \quad (4.21)$$

For the calculation of $f(T)$ one must know the phonon density of states $Z(\Omega)$. In Fig. 4.6 a realistic phonon spectrum is compared with model distributions.

The Debye model often gives a good approximation for the realistic case. In this model the phonon density of states is given by

$$Z(\Omega) = \frac{9N\hbar^3}{k_B^3} \frac{\Omega^2}{\Theta^3} \quad \text{for } \Omega \leq \Omega_D \quad (4.22)$$

For $\Omega > \Omega_D$ one obtains $Z(\Omega) = 0$, where $\Omega_D = k_B \Theta / \hbar$ is the Debye frequency, and Θ is the Debye temperature. Substitution of Eq. (4.22) into Eq. (4.20) gives for the mean square displacement $\langle x^2 \rangle$ in the Debye model

$$\langle x^2 \rangle = \frac{3\hbar^2}{4Mk_B \Theta} \left[1 + 4\left(\frac{T}{\Theta}\right)^2 \int_0^{\Theta/T} \frac{y}{\exp y - 1} \, dy \right] \quad (4.23)$$

where $y = \hbar\Omega/k_B T$. This is shown graphically as a function of temperature in Fig. 4.7.

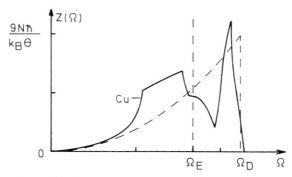

Fig. 4.6 Density of states for the Einstein and Debye model (dashed lines) and for a realistic case (Cu, solid line (Nicklow et al., 1967)).

42 Mössbauer Effect

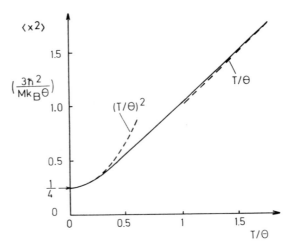

Fig. 4.7 Mean square displacement $\langle x^2 \rangle$ of atoms in the Debye model as a function of temperature.

For high temperatures ($T \gg \Theta$), $\langle x^2 \rangle$ is linear with temperature

$$\langle x^2 \rangle = \frac{3\hbar^2}{Mk_B\Theta}\left(\frac{T}{\Theta}\right) \quad (T \gg \Theta) \tag{4.24}$$

For low temperatures ($T \ll \Theta$) one obtains a T^2 dependence (the integral in Eq. (4.23) is $\pi^2/6$, since $\Theta/T \to \infty$)

$$\langle x^2 \rangle = \frac{3\hbar^2}{Mk_B\Theta}\left[\frac{1}{4} + \frac{\pi^2}{6}\left(\frac{T}{\Theta}\right)^2\right] \quad (T \ll \Theta) \tag{4.25}$$

At low temperatures, the Debye–Waller factor in the Debye model is obtained by substituting Eq. (4.25) into Eq. (4.13)

$$f_D(T) = \exp\left\{-\frac{\hbar^2 k^2}{2M}\frac{3}{2k_B\Theta}\left[1 + \frac{2\pi^2}{3}\left(\frac{T}{\Theta}\right)^2\right]\right\} \tag{4.26}$$

The explicit forms of the Debye–Waller factor given in Eqs. (4.21) and (4.26) show some important relationships:

(a) One can obtain information about the phonon frequency spectrum by measuring the Debye–Waller factor.
(b) The Debye–Waller factor decreases with increasing recoil energy $\hbar^2 k^2/2M$ (M is atomic mass). Since $\hbar k = E_\gamma/c$, E_γ cannot be too large if there is to be a measurable Mössbauer effect. Typical values are $E_\gamma < 100$ keV.

(c) Low temperatures are favorable for observation of a large Mössbauer effect ($T < \Theta$). The maximum Debye–Waller factor occurs at $T = 0$ and is

$$f_D(T = 0) = \exp\left(-\frac{3\hbar^2 k^2}{4 M k_B \Theta}\right) \quad (4.27)$$

4.3 Mössbauer Sources and Experimental Apparatus

4.3.1 Mössbauer Sources

Mössbauer sources must fulfill the following conditions:

(a) The experimentally observed γ transition must lead to the ground state since, for practical purposes, absorption can take place only from the ground state of a sufficiently stable isotope.
(b) The Debye–Waller factor should not be too small. As shown above, favorable conditions are low temperature, low γ energy, high Debye temperature, and large atomic mass.
(c) The lifetime of the Mössbauer level should not be too short. Otherwise the linewidth would be too large and the energy resolution too poor.
(d) Properties of the parent isotope (parent lifetime, chemical and metallurgical properties, etc.) as well as ease of isotope production, determine the practicality of handling and use in the experimental laboratory.

Several Mössbauer sources are discussed in detail below.

^{57}Co–^{57}Fe

The isotope ^{57}Co can be produced by the nuclear reaction $^{56}Fe(d, n)^{57}Co$. The decay of ^{57}Co occurs overwhelmingly (99.8%) by electron capture (EC) from the K shell. The resulting K-shell hole is filled subsequently from a higher shell with the accompanying emission of a 6.4 keV X-ray. The decay scheme is shown in Fig. 4.8.

The Mössbauer source ^{57}Co–^{57}Fe has ideal properties for many applications: The parent isotope ^{57}Co has a sufficiently long halflife ($t_{1/2} = 270$ days) to allow many experiments without changing sources. The lifetime of the Mössbauer level ($t_{1/2} = 98$ ns) leads to a linewidth which has sufficiently good energy resolution for most applications without requiring the extreme experimental measures (see Section 4.3.2) sometimes necessary for sharper lines. Finally, the 14.4 keV energy is sufficiently small that room-temperature measurements are almost always possible with fairly large Debye-

44   Mössbauer Effect

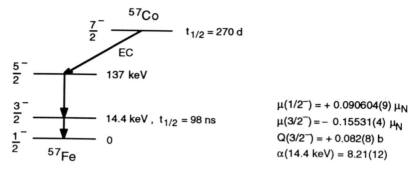

Fig. 4.8 Decay scheme of ^{57}Co. The nuclear moments of the ground state ($I = 1/2$) and the Mössbauer level ($I = 3/2$) are shown along with the total conversion coefficient α of the 14.4 keV transition. All values are from Stevens and Stevens (1977) except for $Q(3/2^-)$ which is from Vajda et al. (1981).

–Waller factors. Smaller γ-ray energies could present experimental difficulties, for example, in discriminating the γ-rays from the accompanying X-rays.

$^{67}Ga-^{67}Zn$

Because of the long lifetime of the Mössbauer level, the ^{67}Zn energy resolution is the best obtainable with practical Mössbauer sources. The very small linewidth of this resonance (corresponding Doppler velocity $\beta = 0.31$ μm s^{-1}) places extreme requirements on the stability of the spectrometer and the preparation of absorber and source. However, the extremely high energy resolution permits extremely precise measurements. The decay scheme is shown in Fig. 4.9.

^{119}Sn–isomer

^{119}Sn is usually made in a reactor by neutron capture from ^{118}Sn. Unlike most other Mössbauer sources, the parent and daughter are the same element. Since the quadrupole moment of the Mössbauer level is relatively small, ^{119}Sn is not well suited for investigation of electric quadrupole interactions. The decay scheme is shown in Fig. 4.10.

$^{151}Sm-^{151}Eu$

^{151}Sm is produced most easily in a reactor by thermal neutron capture by ^{150}Sm. The Mössbauer transition can also be populated by decay of ^{151}Gd. Both decay schemes are shown in Fig. 4.11.

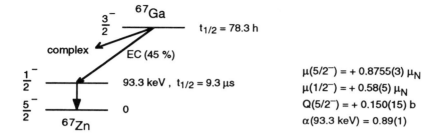

Fig. 4.9 Portion of the decay scheme of ^{67}Ga. For clarity, only the most abundant decay mode (45%) is shown. The nuclear moments of the ground state ($I = 5/2$) and the Mössbauer level ($I = 1/2$) are shown on the right side along with the total conversion coefficient α of the 93.3 keV transition. Values are from Stevens and Stevens (1977) and Lederer and Shirley (1978).

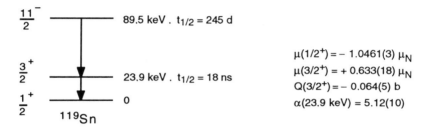

Fig. 4.10 Decay scheme of the isomeric level ^{119m}Sn. The nuclear moments of the ground state ($I = 1/2$) and the Mössbauer level ($I = 3/2$) are shown on the right side of the figure along with the total conversion coefficient α of the 23.9 keV transition. Values are from Stevens and Stevens (1977).

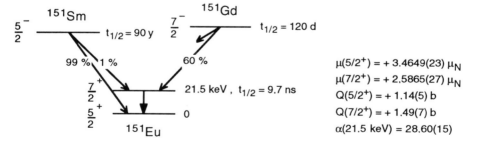

Fig. 4.11 Decay scheme of ^{151}Sm and ^{151}Gd. The nuclear moments of the ground state ($I = 5/2$) and the Mössbauer level ($I = 7/2$) are shown on the right side of the figure along with the total conversion coefficient α of the 21.5 keV transition. Values are from Stevens and Stevens (1977).

4.3.2 Mössbauer Apparatus

A schematic representation of a Mössbauer spectrometer is shown in Fig. 4.12. It consists of a radioactive source which can be moved, an absorber, and a detector for the γ-rays.

In principle one can distinguish two different arrangements: (a) transmission geometry, in which the intensity of the transmitted γ-rays is measured, and (b) scattering geometry, in which the fluorescence radiation is measured. In transmission geometry, one obtains reduced transmission, and in reflection geometry enhanced fluorescence at resonance. In the reflection geometry, one can detect either X-rays or conversion electrons (see Figs. 4.8–4.11 for conversion electron coefficients) instead of the γ-rays. X-rays are emitted following the conversion process when the resulting inner-shell hole is refilled. Both the X-ray and conversion electron production rates reflect increased absorption at resonance. Except when noted, we will limit discussion to transmission geometry.

One uses the Doppler effect to achieve the resonance condition, i.e. one moves the source relative to the absorber or vice versa. An electromagnetic drive based on the loudspeaker principle is often used for the movement. The source (or absorber) is part of a mechanical oscillator system (see Fig. 4.13) on which a certain velocity profile can be imposed by means of a driver coil. The most common velocity profiles are:

(a) triangular or sawtooth form,
(b) sine wave (used when high velocities or high stability are required),
(c) constant velocity.

The velocity profile is imposed by a function generator. The velocity is detected by the pickup coil and fed into the feedback amplifier which provides an error signal that keeps the velocity close to the desired

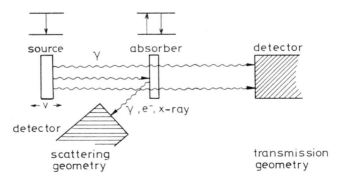

Fig. 4.12 Schematic representation of a Mössbauer spectrometer.

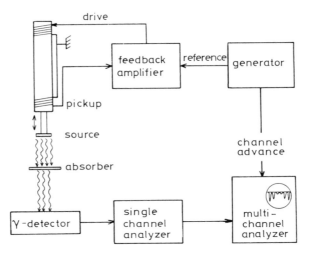

Fig. 4.13 Block diagram of a Mössbauer spectrometer.

waveform. Typically the actual velocity profile differs from the ideal profile by less than 0.1%.

The γ-ray detector measures the spectrum of the γ-rays transmitted by the absorber. Normally, either thin scintillation detectors or gas-filled proportional counters are used. The pulse height spectrum of ^{57}Co measured with a proportional counter filled with 97% Kr and 3% CO_2 is shown in Fig. 4.14. One recognizes the 14.4 keV Mössbauer line and the 6.4 keV X-ray line of ^{57}Fe. The line at 1.8 keV originates from the 14.4 keV quantum which has knocked out a Kr K-shell electron, thereby losing 12.6 keV energy. Both the 14.4 keV and 1.8 keV lines can be used to detect absorption of the 14.4 keV γ-ray.

The single-channel analyzer 'window' is set so that only signals with pulse height within the Mössbauer line are recorded in the multichannel analyzer. The multichannel analyzer is used in the multi-count mode. In this mode, counts originating from the single-channel analyzer are added into a certain channel until the next channel is selected. Then counts are collected in the next channel, etc. The channel selection is controlled by the function generator so that during a period of the velocity waveform all channels are selected for equal time intervals. At the beginning of a new period, the multichannel analyzer is switched back to the first channel. This synchronization assures that each channel is associated with a certain drive velocity. For triangular or sawtooth waveforms, the channel number is directly proportional to the velocity. For a sine wave velocity profile, a simple correction is required to obtain a spectrum of count rate vs. velocity. This is usually done by a computer.

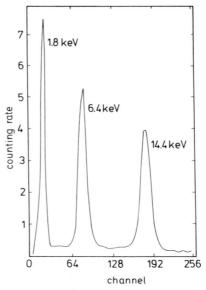

Fig. 4.14 Pulse height spectrum of a ^{57}Co Mössbauer source measured by a proportional counter filled with 97% Kr and 3% CO_2. The high-energy γ-quanta (123 keV and 137 keV) are not shown.

4.4 Isomer Shift

We have seen that the energy of a nucleus with mean square radius $\langle r^2 \rangle$ is shifted relative to a point nucleus. This energy is described by the monopole term as shown in Eq. (3.26),

$$E_C = \frac{Ze^2}{6\varepsilon_0} |\psi(0)|^2 \langle r^2 \rangle \qquad (4.28)$$

In a Mössbauer experiment, differences in the monopole energies cause a shift in the resonance. For a moving source S, the energy of the γ-ray due to a transition from the excited state (index e) to the ground state (index g) is

$$\hbar\omega(S) = \hbar\omega_0 \left(1 + \frac{v}{c}\right) + \frac{Ze^2}{6\varepsilon_0} |\psi_S(0)|^2 (\langle r_e^2 \rangle - \langle r_g^2 \rangle) \qquad (4.29)$$

The first term is the transition energy modified by the Doppler shift due to source motion, and the second term is the monopole energy of the nucleus in the excited state minus that in the ground state. For a stationary absorber A, the energy needed to excite the nucleus from its ground state to the

Mössbauer state is

$$\hbar\omega(A) = \hbar\omega_0 + \frac{Ze^2}{6\varepsilon_0}|\psi_A(0)|^2(\langle r_e^2\rangle - \langle r_g^2\rangle) \qquad (4.30)$$

Resonant absorption occurs if $\omega(S) = \omega(A)$. In that case, the velocity of the source is

$$v_{\text{res}} = \frac{Ze^2c}{6\varepsilon_0\hbar\omega_0}[|\psi_A(0)|^2 - |\psi_S(0)|^2](\langle r_e^2\rangle - \langle r_g^2\rangle) \qquad (4.31)$$

v_{res} is called the isomer shift; it is often designated S in Mössbauer literature. The isomer shift sign convention is that v_{res} is positive when the source is moving toward the absorber.

From Eq. (4.31) one sees that the isomer shift measures energy differences only. One must always specify the reference, e.g. isomer shift of Fe in Fe compared to Fe in Pt. For a nonzero isomer shift to occur, the following conditions must be fulfilled simultaneously:

$$|\psi_A(0)|^2 - |\psi_S(0)|^2 \neq 0 \quad \text{and} \quad \langle r_e^2\rangle - \langle r_g^2\rangle \neq 0 \qquad (4.32)$$

For ^{57}Fe, $\langle r_e^2\rangle - \langle r_g^2\rangle$ is negative, i.e. the nucleus has a smaller radius in the excited state than in the ground state. The nuclear radius of ^{119}Sn and ^{151}Eu is larger in the excited state than in the ground state, so $\langle r_e^2\rangle - \langle r_g^2\rangle$ is positive.

Fig. 4.15 shows a measurement of the isomer shift for ^{119}Sn. BaSnO$_3$ was used as source, and the absorber was β-Sn (metallic white tin). The resonance line in the upper spectrum is unsplit but shifted relative to $v = 0$. The isomer shift is $S = 2.56$ mm s^{-1}. In the lower spectrum, conversion electrons were detected instead of transmitted γ-rays.

Conversion electron Mössbauer spectroscopy (CEMS) offers interesting possibilities for application in surface physics. Conversion electrons have low energy and can escape only from a depth of order 100 nm or less, consequently providing information only about the near-surface region. The energy loss of the emitted conversion electron is proportional to the escape depth. Therefore a measurement of the Mössbauer absorption as a function of conversion electron energy provides depth-dependent information (depth resolution is approximately 5 nm).

The surface sensitivity of CEMS is illustrated in the lower spectrum of Fig. 4.15. The small additional line at $v = 0$ arises from absorption in an approximately 3 nm surface layer of SnO$_2$ on the tin absorber. The corresponding line in transmission geometry is not visible since absorption in the surface layer is negligible compared to the total absorption ($d = 0.125$ mm for the absorber).

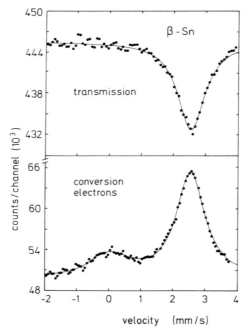

Fig. 4.15 Mössbauer spectrum for ^{119}Sn in white tin (metallic β-Sn). The upper spectrum was obtained in transmission geometry by γ-ray detection, the lower one in reflection geometry by detecting conversion electrons. The source was BaSnO$_3$. The line at $v = 0$ in the lower spectrum arises from a thin SnO$_2$ surface layer on the absorber.

4.4.1 Isomer Shift and Chemical Valence

A close relationship between isomer shift and chemical valence has been observed for many Mössbauer atoms. This correlation is discussed here for tin compounds (Fig. 4.16).

Compounds containing divalent tin have isomer shifts between $+2.3$ and $+4.4$ mm s^{-1}. Compounds with tetravalent tin have shifts around $+2$ mm s^{-1} if the bonding is strongly covalent (e.g. α-Sn) and around zero if the bonding is strongly ionic (e.g. K$_2$SnF$_6$). All shifts are measured relative to BaSnO$_3$. This qualitative dependence is attributed to the influence of the 5s electrons. Neutral Sn has the electron configuration

$$\text{Sn:} \ldots (4d)^{10}(5s)^2(5p)^2$$

In Sn(IV)-salts (ionic bonding) the four 5s and 5p electrons are transferred to the anion, i.e. tin has the configuration Sn^{4+}: ... $(4d)^{10}$, so there are no 5s electrons at the Sn nucleus. For Sn(II) salts, the two 5s electrons remain

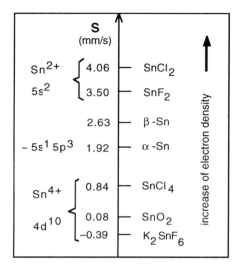

Fig. 4.16 Isomer shifts for α-Sn, β-Sn, and some Sn compounds. The ionization state and electron configurations are shown on the left side. Source: BaSnO₃. Taken from Flinn (1978).

bound to Sn and therefore cause a large positive isomer shift. For covalent tetravalent Sn compounds approximately one 5s electron remains bound to the Sn atom (sp³ hybridization similar to carbon). Since the isomer shift depends on the electron density at the nucleus one expects in first order a linear relation between the isomer shift and the number of 5s electrons on the tin atom. This qualitative dependence is experimentally verified, but a number of corrections need to be considered for a quantitative understanding.

The strongest correction is caused by 5p electrons whose shielding reduces the nuclear potential experienced by 5s electrons and consequently leads to weaker 5s binding and reduction of the 5s density at the nucleus. Thus 5p electrons reduce the isomer shift. The following empirical relation has been found experimentally (Flinn, 1978):

$$S = -0.38 + 3.10 n_s - 0.20 n_s^2 - 0.17 n_s n_p \qquad (4.33)$$

where n_s is the number of 5s electrons and n_p the number of 5p electrons bound to Sn; n_s and n_p are effective occupation numbers and can assume non-integral values. The constant term arises because Sn in the reference BaSnO₃ does not have a completely ionic $(3d)^{10}$ outer-shell configuration (the $(3d)^{10}$ outer shell is a better approximation in K₂SnF₆). The second term in Eq. (4.33) corresponds to the linear dependence of the isomer shift on n_s as discussed above. The third term is a correction not discussed here, and the fourth term is due to the shielding effect of 5p electrons.

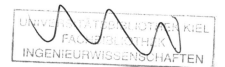

4.4.2 Valence Fluctuations

In the preceding section we showed that the isomer shift of ^{119}Sn depends strongly on the valence of the Sn. The same is true for the Mössbauer atom ^{151}Eu, which has isomer shifts (relative to Eu_2O_3) around -11 mm s^{-1} in the divalent state (Eu^{2+}) and around zero in the trivalent state (Eu^{3+}). One finds isomer shift values that are intermediate between these two values and are temperature dependent for some Eu compounds. As an example, a measurement on $EuCu_2Si_2$ is shown in Fig. 4.17. In the upper part of the figure, a spectrum of ^{151}Eu in $EuAg_2Si_2$ and in the lower part ^{151}Eu in $EuFe_2Si_2$ is shown. Eu is divalent in the first and trivalent in the second material. One sees clearly that the isomer shifts of the spectra for Eu in $EuCu_2Si_2$ lie between those of Eu^{2+} and Eu^{3+} and are temperature dependent (Fig. 4.18). This observation indicates that Eu in $EuCu_2Si_2$ has an intermediate valence.

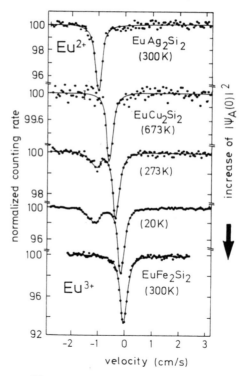

Fig. 4.17 Isomer shift of ^{151}Eu in $EuCu_2Si_2$ compared to Eu_2O_3. For comparison, the spectrum of Eu^{2+} in $EuAg_2Si_2$ and in the lower part the spectrum of Eu^{3+} in $EuFe_2Si_2$ is shown. Taken from Bauminger et al. (1973).

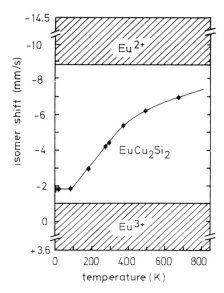

Fig. 4.18 Isomer shift of ^{151}Eu in EuCu$_2$Si$_2$ compared with Eu$_2$O$_3$ as a function of temperature. The experimentally measured region of the isomer shift in Eu^{2+} and Eu^{3+} is shown for comparison (Bauminger et al., 1973).

The authors interpret their data in the following way: Eu^{2+} has the configuration (4f)^7 having seven electrons in the 4f shell, and Eu^{3+} has the configuration (4f)^6 with six electrons in the 4f shell. It is assumed that the seventh electron is neither permanently bound nor permanently free but that it changes rapidly ($t < 3.5 \times 10^{-11}$ s) between the localized 4f level and the conduction band. This effect is called a valence fluctuation. One only observes the average isomer shift since many valence fluctuations occur during the lifetime of the Mössbauer state. The temperature dependence is attributed to an increase with temperature of the probability that the seventh 4f electron is bound to Eu.

Valence fluctuations are also observed for other compounds of rare-earth elements. There is presently no fully satisfactory quantitative explanation of the valence fluctuation phenomenon.

4.5 Electric Quadrupole Interaction

As discussed in Chapter 3, the degeneracy of the M sublevels is lifted if a nonspherical nucleus interacts with an electric field gradient. The energy

54   Mössbauer Effect

level splitting does not depend on the sign of M; one cannot distinguish one end of a cigar-shaped distribution from the other. One has (Eqs. (3.39) and (3.41))

$$E_Q = [3M^2 - I(I+1)]\hbar\omega_Q \tag{4.34}$$

As discussed previously, there is no splitting if the nuclear spin $I < 1$.

We will first study the electric quadrupole splitting for the Mössbauer isotope ^{57}Fe. Fig. 4.19 shows the level splitting of ^{57}Fe due to a quadrupole interaction. The energy separation of the two Mössbauer lines is

$$\Delta E = 6\hbar\omega_Q = \hbar\omega_0 \frac{\Delta v}{c} \tag{4.35}$$

It follows that the separation of the two lines in the velocity spectrum is

$$\Delta v = \frac{eQV_{zz}c}{2\hbar\omega_0} \tag{4.36}$$

The experimental result for $FeSO_4 \cdot 7H_2O$ is shown in Fig. 4.20. The observed quadrupole splitting $\Delta v = 3.16(1)$ mm s^{-1} (the average value is nonzero because of an isomer shift) is qualitatively understandable in the following way: In $FeSO_4$, iron is divalent, i.e. it has six 3d electrons. The half-filled 3d shell $(3d)^5$ is spherically symmetric and makes no contribution to the electric field gradient. The observed field gradient therefore is largely due to the sixth 3d electron. As expected from this reasoning, trivalent iron salts, which have closed Fe 3d shells, have weak or undetectable quadrupole splittings.

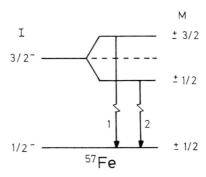

Fig. 4.19   Quadrupole splitting of the ^{57}Fe Mössbauer transition.

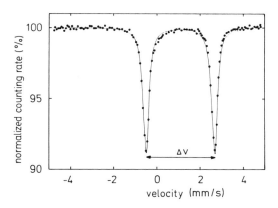

Fig. 4.20 The ^{57}Fe Mössbauer spectrum of $FeSO_4 \cdot 7H_2O$. The source was ^{57}Co in Rh metal (single line source).

Line intensities

One finds experimentally that the intensities of the two quadrupole-split Mössbauer lines of ^{57}Fe are equal in powder samples. We will present a brief explanation. The transition probability of a transition from the excited state $|3/2, M_e\rangle$ to the ground state $|1/2, M_g\rangle$ by emission of pure M1 radiation is given by

$$I_\gamma \propto |\langle 1/2, M_g|\mathcal{M}(\text{M1})|3/2, M_e\rangle|^2 F_{lm}(\theta) \tag{4.37}$$

The first term is the absolute square of the nuclear matrix element of the M1 multipole operator $\mathcal{M}(\text{M1})$, while the second describes the radiation emission pattern (see Eq. (2.36)). By using the Wigner–Eckart theorem and neglecting the constant reduced matrix element, one obtains

$$I_\gamma \propto \begin{pmatrix} 1/2 & 1 & 3/2 \\ -M_g & m & M_e \end{pmatrix}^2 F_{lm}(\theta) \tag{4.38}$$

The 3j-symbols appearing in Eq. (4.38) have the following values (Rotenberg *et al.*, 1959):

$$\begin{pmatrix} 1/2 & 1 & 3/2 \\ \mp 1/2 & \mp 1 & \pm 3/2 \end{pmatrix}^2 = \frac{3}{12}$$

$$\begin{pmatrix} 1/2 & 1 & 3/2 \\ \mp 1/2 & 0 & \pm 1/2 \end{pmatrix}^2 = \frac{2}{12} \tag{4.39}$$

$$\begin{pmatrix} 1/2 & 1 & 3/2 \\ \mp 1/2 & \mp 1 & \pm 1/2 \end{pmatrix}^2 = \frac{1}{12}$$

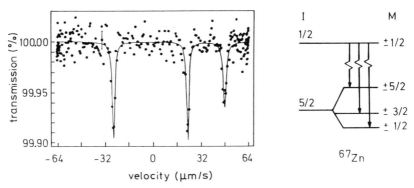

Fig. 4.21 The ^{67}Zn Mössbauer spectrum of zinc metal. The source was ^{67}Ga in Cu (single line source). Taken from Potzel et al. (1978). The quadrupole splitting of the relevant ^{67}Zn levels is shown on the right side.

In powder samples, the orientation of the electric field gradients are statistically distributed, i.e. the γ-rays are isotropically emitted. Then one must take the angular average of $F_{lm}(\theta)$, which is unity. The line intensities are therefore proportional to the squares of the 3j-symbols. One obtains for the individual transitions in Fig. 4.19: $I_1 \propto 3/12$, $I_2 \propto 2/12 + 1/12 = 3/12$. Thus, $I_1 = I_2$, in agreement with experimental observation.

As another example of quadrupole splitting, we describe ^{67}Zn Mössbauer experiments. Fig. 4.21 shows the spectrum for a metallic polycrystalline ^{67}Zn absorber and a ^{67}Ga source in Cu (single line source) along with the level splitting diagram for this isotope. The electric field gradient arises from the non-cubic (hexagonal) crystal structure of zinc (see Section 5.5). Note the extraordinarily high resolution (velocity axis in μm s^{-1}) which allows the measurement of this very weak quadrupole splitting.

4.6 Magnetic Dipole Interaction

The magnetic interaction $-\boldsymbol{\mu} \cdot \boldsymbol{B}$ leads to nuclear level splitting with equal separation between magnetic sublevels (Eq. (3.3)) and consequently to a splitting in the energy of emitted γ-rays. We discuss again the example of ^{57}Fe. Fig. 4.22 shows the magnetic splitting for ^{57}Fe.

One observes experimentally only six transitions, and these are the ones allowed by M1 selection rules. It follows that the radiation is pure M1. The transition energies are

$$\hbar\omega(M_e \to M_g) = \left(E_e - \frac{\mu_e}{I_e}M_e B\right) - \left(E_g - \frac{\mu_g}{I_g}M_g B\right)$$

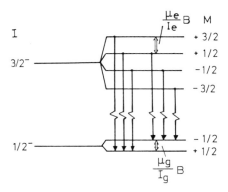

Fig. 4.22 Splitting of the nuclear levels of ^{57}Fe in a B field. The γ-ray transitions shown in the figure obey the M1 radiation selection rules: $m = 0, \pm 1$.

$$= \hbar\omega_0 - \left(\frac{\mu_e}{I_e}M_e - \frac{\mu_g}{I_g}M_g\right)B \qquad (4.40)$$

The resonance velocities for the case of a moving source are obtained from the following conditions,

$$\hbar\omega_0\left(1 + \frac{v_{\text{res}}}{c}\right) = \hbar\omega_0 - \left(\frac{\mu_e}{I_e}M_e - \frac{\mu_g}{I_g}M_g\right)B \qquad (4.41)$$

which yield the resonance velocities,

$$v_{\text{res}} = -\frac{c}{\hbar\omega_0}\left(\frac{\mu_e}{I_e}M_e - \frac{\mu_g}{I_g}M_g\right)B \qquad (4.42)$$

4.6.1 Magnetic Hyperfine Field in Bulk Iron

^{57}Fe is an ideal probe for measuring the magnetic hyperfine fields B_{hf} at substitutional lattice sites and thus to gain insight into ferromagnetism. Fig. 4.23a shows the Mössbauer spectrum of ^{57}Fe in metallic Fe for the single line source ^{57}Co in Pt. One sees the splitting of the Mössbauer line into the six-line spectrum as discussed above. This measurement has been used to determine the B_{hf} field at the nucleus and the magnetic moment μ_e of the excited state (μ_g is known from other measurements). As a result, one finds at 4 K

$$\mu_e = -0.153(4)\ \mu_N \quad \text{and} \quad B = -33.3(10)\ \text{T} \qquad (4.43)$$

58   Mössbauer Effect

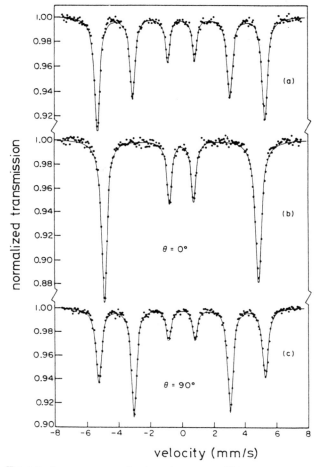

Fig. 4.23 The ^{57}Fe Mössbauer spectrum for iron. Source was ^{57}Co in Pt. In (b) and (c) the line splitting is somewhat changed due to the external magnetic field used to magnetize the sample. (a) Unmagnetized iron absorber (directions of internal B fields are statistically distributed). (b) Magnetized iron as absorber with magnetization parallel to the γ-ray emission direction k. (c) Magnetized iron as absorber with magnetization perpendicular to the γ-ray emission direction k (Gonsor, 1975).

The large negative local B field is due to polarization of the s electrons (Fermi contact interaction, see Section 6.5) by the 3d electrons. The negative sign means that the B_{hf} field and the external magnetization of the sample have opposite orientations. The dependence of the local field on temperature closely follows the temperature dependence of the macroscopic magnetization.

Line intensity

Line intensities of unmagnetized samples (statistically distributed local field directions) can be derived from the $3j$-symbols (Eq. 4.39) and are found to be in the ratio $1:2:3$ (see Fig. 4.23a).

For magnetized samples, the line intensities depend on the relative direction of the magnetization with respect to the γ-ray emission direction. We discuss two special cases:

(a) ***B*** *field parallel* to γ-ray emission direction. The z axis is again taken as the direction of the ***B*** field. The angular dependent factor in Eq. (4.37) has the value

$$F_{10}(\theta = 0°) = 0 \quad \text{and} \quad F_{1\pm1}(\theta = 0°) = 1/2 \qquad (4.44)$$

This means that lines with $m = 0$ are not observed. One obtains an intensity ratio $1:0:3$ (see Fig. 4.23b).

(b) ***B*** *field perpendicular* to γ-ray emission direction. The angular dependent part is now

$$F_{10}(\theta = 90°) = 1/2 \quad \text{and} \quad F_{1\pm1}(\theta = 90°) = 1/4 \qquad (4.45)$$

Using the appropriate $3j$-symbols, this leads to an intensity ratio $1:4:3$ (see Fig. 4.23c).

4.6.2 Magnetic Hyperfine Field at the (110) Surface of Iron

As we have seen, the local ***B*** field in the interior of iron is equal at all lattice sites. It is interesting to know whether this is also true for atoms at the surface. Gradmann and collaborators (Korecki and Gradmann, 1985) have investigated the (110) surface of single-crystal iron films using conversion electron Mössbauer spectroscopy (CEMS, see Section 4.4). The films were prepared by evaporation of iron in ultrahigh vacuum ($p \approx 3 \times 10^{-9}$ Pa) on a (110) W substrate. First they evaporated a film of nonresonant ^{56}Fe, then a monolayer of the Mössbauer isotope ^{57}Fe. To investigate layers below the surface they evaporated additional ^{56}Fe. Fig. 4.24 shows the Mössbauer spectrum for a ^{57}Fe monolayer at the surface. The sample layer structure is illustrated schematically in the figure. A typical magnetic splitting is clearly observed, with the second and fifth lines of the six-line spectrum being very weak. In the geometry used in this experiment (angle $\theta = 15°$ between ***B*** and ***k***) the observed line intensities prove that the local ***B*** field lies parallel to the surface and coincides with a $\langle 110 \rangle$ direction.

The dependence of the local ***B*** field on the distance of the ^{57}Fe monolayer

60   Mössbauer Effect

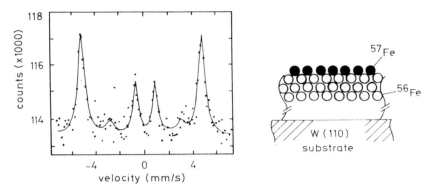

Fig. 4.24 Mössbauer spectrum for ^{57}Fe on a (110) iron surface (Korecki and Gradmann, 1985). The sample layer structure is shown schematically on the right.

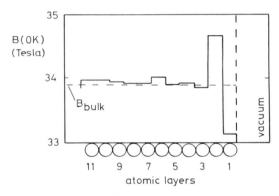

Fig. 4.25 Magnetic hyperfine field at ^{57}Fe, extrapolated to $T = 0$ as a function of the monolayer distance from the (110) surface. Taken from Korecki and Gradmann (1986).

from the surface is shown in Fig. 4.25. The data shown in this figure are the extrapolation of the **B** fields to $T = 0$ from room temperature measurements, assuming a $T^{3/2}$ dependence (Bloch spin-wave behavior). One sees clearly that the magnetic hyperfine field in the surface layer is reduced and in the layer below enhanced relative to the bulk value. The authors attribute this oscillatory behavior to the influence of conduction electron Friedel oscillations near the surface.

4.7 Quadratic Doppler Effect

Thermal motion of atoms gives rise to a frequency shift of emitted γ-rays due to the quadratic Doppler effect. The rigorous relativistic formulation of the Doppler effect yields

$$\omega = \omega_0 \sqrt{1 - \frac{v^2}{c^2}} + \mathbf{k} \cdot \mathbf{v} = \omega_0 + \mathbf{k} \cdot \mathbf{v} - \frac{\omega_0}{2}\frac{v^2}{c^2} + \cdots \quad (4.46)$$

where $\mathbf{k}$ ($|\mathbf{k}| = \omega/c$) and ω are the wave vector and the angular frequency of the γ-ray in the rest frame of the observer; ω_0 is the angular frequency in the rest frame of the emitter. For $\mathbf{v}$ in the direction of $\mathbf{k}$ one obtains: $\omega(1 - v/c) = \omega_0\sqrt{1 - v^2/c^2}$.

The term $\mathbf{k} \cdot \mathbf{v}$ in Eq. (4.46), due to the linear Doppler effect, was discussed above (Eq. (4.6)). We will now discuss the higher-order corrections. In a crystal lattice the emitter atom will oscillate many times during the lifetime of the Mössbauer level ($t_{latt} \approx 10^{-13}$ s, $t_N \approx 10^{-7}$ s). The average velocity v_{latt} is zero, so the linear Doppler shift terms average to zero. However, the quadratic terms do not average to zero. The time-average emission frequency from the locally vibrating source is

$$\bar{\omega} = \omega_0 - \frac{\omega_0}{2c^2}\overline{(v_{latt})^2} = \omega_0 - \frac{\omega_0}{Mc^2}\frac{\overline{M(v_{latt})^2}}{2} \quad (4.47)$$

where $(1/2)\overline{M(v_{latt})^2}$ is the time average kinetic energy of an atom. For a harmonic oscillator, the kinetic energy is on average half the total energy. With the molar energy E_{mole} and

$$E_{mole} = 2N_A \frac{\overline{M(v_{latt})^2}}{2} \quad (4.48)$$

one obtains

$$\bar{\omega} = \omega_0 - \omega_0 \frac{E_{mole}}{Mc^2 2N_A} \quad (4.49)$$

where N_A is Avogadro's constant.

By differentiating with respect to temperature and using the expression for the molar specific heat $C_{mole} = dE_{mole}/dT$, one obtains the frequency shift per degree,

$$\frac{d\bar{\omega}}{dT} = -\omega_0 \frac{C_{mole}(T)}{Mc^2 2N_A} \quad (4.50)$$

We will estimate the frequency shift due to the quadratic Doppler effect

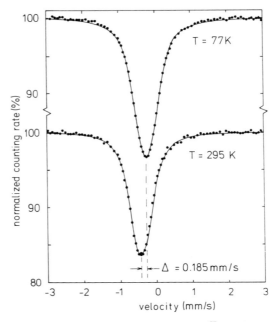

Fig. 4.26 Quadratic Doppler effect for the 14.4 keV line of ^{57}Fe. The spectra for a stainless steel absorber at 77 K and at 295 K are shown. The source (^{57}Co in Rh) was held at 295 K in both cases.

for ^{57}Fe. At high temperature, Dulong and Petit's law gives $C_{\text{mole}} = 3R = 24.9 \, \text{J mol K}^{-1}$, where R is the gas constant. Inserting this value in Eq. (4.50) yields for ^{57}Fe

$$-\frac{d\bar{\omega}}{dT}\frac{1}{\omega_0} = 2.4 \times 10^{-15} \, \text{K}^{-1} \tag{4.51}$$

For an experiment in which the absorber temperature is changed from 77 K to 295 K, one obtains a relative frequency shift of $\Delta\omega_0/\omega_0 \approx 5 \times 10^{-13}$. This value is the same order of magnitude as the natural linewidth and is consequently readily measurable. Fig. 4.26 shows such a measurement for ^{57}Fe in nonmagnetic stainless steel. The experimentally observed linewidth is approximately four times larger than the natural linewidth of ^{57}Fe because of the different lattice sites in stainless steel (nonresolved quadrupole splittings). The experimental resonance shift between 77 K and 295 K is $\Delta = 0.185(1) \, \text{mm s}^{-1}$ corresponding to a relative frequency shift of 6.2×10^{-13}. This compares well with the estimate made above.

5
Perturbed γ–γ Angular Correlation (PAC)

In contrast to the Mössbauer effect, the perturbed γ–γ angular correlation method (PAC) measures the time dependence of the γ-ray emission pattern, not the nuclear level energy splitting. One can also consider PAC to be a measurement of the rotation or precession of the angular correlation. The origin of the precession is the hyperfine interaction.

The basic requirement for PAC is that the γ-rays are emitted anisotropically from the probe nucleus. Otherwise no precession would be observable. An anisotropic angular distribution is obtained only if the state from which the radiation is emitted is at least partly polarized or aligned, i.e. the M substates are not equally populated. The non-equal M-state population can be obtained, for example, by coincident observation of γ-rays emitted when the nuclear level is populated following decay from a higher level. Because of the importance of the anisotropic angular distribution, we will discuss first unperturbed γ–γ angular correlation. We will then discuss the perturbation due to the hyperfine interaction in the intermediate state. A rigorous presentation of PAC is given by Frauenfelder and Steffen (1965).

5.1 Theory of Unperturbed γ–γ Angular Correlations

The general theory of angular correlations is mathematically quite complex, and the physical concepts are not always obvious from the mathematics. Therefore we will present a simpler version, the so-called naive theory, and will then only briefly sketch the general theory before introducing the perturbations.

A schematic illustration of an angular correlation measurement is shown in Fig. 5.1. The initial state $|I_i, M_i\rangle$ decays by emission of γ_1 into the intermediate state $|I, M\rangle$ and then into the final state $|I_f, M_f\rangle$ by emission of γ_2. Emission γ_1, having multipolarity l_1 and projection m_1, is detected in

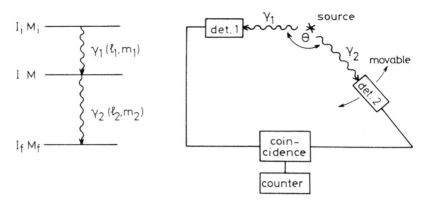

Fig. 5.1 Illustration of the γ–γ angular correlation apparatus.

the spatially fixed detector (det.1), and γ_2, having quantum numbers l_2 and m_2, in the movable detector (det.2). The coincident count rate as a function of angle θ gives the angular correlation.

5.1.1 Naive Theory

Assuming equal M-state occupation in the initial nuclear state (I_i, M_i in Fig. 5.1), the population of an M substate in the intermediate state is given by

$$P(M) = \sum_{M_i} G(M_i \rightarrow M) F_{l_1 m_1}(\theta_1) \tag{5.1}$$

where $F_{l_1 m_1}(\theta_1)$ is the angular distribution function (see Eq. (2.36)) of γ-quanta having multipolarity l_1 and $m_1 = M_i - M$. $F_{l_1 m_1}(\theta_1)$ thus gives the probability that γ-rays will be emitted at angle θ_1 with respect to the z axis.

The factor $G(M_i \rightarrow M)$, the integrated rate for the transition between $|I_i, M_i\rangle$ and $|I, M\rangle$, is given by

$$G(M_i \rightarrow M) = |\langle I, M | \mathcal{M}_{l_1 m_1} | I_i, M_i \rangle|^2 = |\langle I_i, M_i | \mathcal{M}_{l_1 m_1} | I, M \rangle|^2$$

$$= \begin{pmatrix} I_i & l_1 & I \\ -M_i & m_1 & M \end{pmatrix}^2 \langle I \| \mathcal{M}_{l_1} \| I_i \rangle^2 \tag{5.2}$$

where $\mathcal{M}_{l_1 m_1}$ is the multipole operator for multipole radiation with l_1 and m_1. The second line in Eq. (5.2) arises from application of the Wigner–Eckart theorem (Appendix A.3). The reduced matrix element $\langle I \| \mathcal{M}_{l_1} \| I_i \rangle$ is independent of the magnetic quantum numbers and is the same for all transitions between $|I_i, M_i\rangle$ and $|I, M\rangle$. Therefore the relative transition probabilities are given by the squares of the 3*j*-symbols

$$G(M_i \to M) \propto \begin{pmatrix} I_i & l_1 & I \\ -M_i & m_1 & M \end{pmatrix}^2 \tag{5.3}$$

Up to this point the z axis has been arbitrary, but we will now choose a specific axis to simplify further calculations. We choose the z axis as the source to detector 1 direction (emission direction of γ_1). Then $\theta_1 = 0$, and

$$P(M) \propto \sum_{M_i} \begin{pmatrix} I_i & l_1 & I \\ -M_i & \pm 1 & M \end{pmatrix}^2 F_{l_1 \pm 1}(0) \tag{5.4}$$

Because of this special choice of axes, m_1 is limited to the values ± 1, since γ-ray emission in the z direction is nonzero only for $m_1 = \pm 1$ (compare with Fig. 2.6). Thus we see that observation of γ_1 implies that the magnetic substates of the intermediate state $|I, M\rangle$ are not equally populated, since only transitions with $m_1 = \pm 1$ are possible, i.e. $P(M) \neq P(M')$ in general. This implies that the intermediate state can be aligned, but one can show that the state is not polarized, because $P(M) = P(-M)$. Alignment of the intermediate state has the consequence that the emission pattern of the second γ-ray is not isotropic. Since it is the detection of γ_1 that determines the intermediate state alignment, it is important that both γ_1 and γ_2 are measured in coincidence. The emission of γ_2 in coincidence with γ_1 has the following angular dependence (compare with Eq. (5.1))

$$W(\theta) \propto \sum_{M, M_f} P(M) G(M \to M_f) F_{l_2 m_2}(\theta) \tag{5.5}$$

For the second γ-ray emission, we no longer have equal $P(M)$. In Eq. (5.5), θ is the angle between the emission direction of γ_2 and γ_1. Substituting $P(M)$ from Eq. (5.4) and using the Wigner–Eckart theorem for $G(M \to M_f)$ yields

$$W(\theta) \propto \sum_{M_i, M, M_f} \begin{pmatrix} I_i & l_1 & I \\ -M_i & \pm 1 & M \end{pmatrix}^2 F_{l_1 \pm 1}(0) \begin{pmatrix} I & l_2 & I_f \\ -M & m_2 & M_f \end{pmatrix}^2 F_{l_2 m_2}(\theta) \tag{5.6}$$

As demonstrated in the following two examples, this formula allows an easy explicit calculation of the angular correlation.

EXAMPLE 0–1–0 cascade

We first discuss the hypothetical example of a 0–1–0 γ–γ cascade (Fig. 5.2). The nuclear spins $I_i = 0$, $I = 1$, $I_f = 0$ allow only pure dipole transitions, so $l_1 = l_2 = 1$.

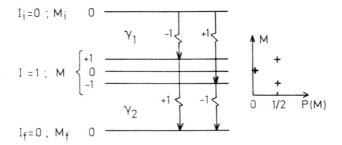

Fig. 5.2 Level scheme for a 0–1–0 γ–γ cascade. The population probabilities of the sublevels in the intermediate state are shown on the right. The M levels are degenerate in energy but are shown separated for clarity.

By choosing the z axis in the emission direction of γ_1, there are only transitions from $|I_i, M_i\rangle$ to $|I, M\rangle$ with $m_1 = \pm 1$; $m_1 = 0$ is not allowed (compare Eq. (5.4)). This has the consequence that $M = 0$ in the intermediate state is not populated. Thus we obtain an alignment in the intermediate state. The subsequent emission of the second γ-ray in coincidence with γ_1 is not isotropic. Substitution in Eq. (5.6) gives

$$W(\theta) \propto \begin{pmatrix} 0 & 1 & 1 \\ 0 & 1 & -1 \end{pmatrix}^2 F_{11}(0) \begin{pmatrix} 1 & 1 & 0 \\ 1 & -1 & 0 \end{pmatrix}^2 F_{1-1}(\theta)$$

$$+ \begin{pmatrix} 0 & 1 & 1 \\ 0 & -1 & 1 \end{pmatrix}^2 F_{1-1}(0) \begin{pmatrix} 1 & 1 & 0 \\ -1 & 1 & 0 \end{pmatrix}^2 F_{11}(\theta) \quad (5.7)$$

Apart from sign, all $3j$-symbols are equal, since any may be transformed into another by exchange of columns (see Appendix A.1). Since they appear quadratically, they may be omitted from the proportionality (5.7). Since $F_{11}(0) = F_{1-1}(0) = 1/2$, we obtain finally

$$W(\theta) \propto F_{1\pm 1}(\theta) \propto 1 + \cos^2 \theta \quad (5.8)$$

EXAMPLE Angular correlation for ^{60}Co

As a second example, we consider the γ–γ cascade in ^{60}Ni, which is often used as a demonstration of unperturbed angular correlation. The γ–γ cascade of ^{60}Ni follows β^--decay of ^{60}Co (see Fig. 5.3).

The transitions in ^{60}Ni have pure E2 character, so Eq. (5.6) applies

Theory of Unperturbed γ–γ Angular Correlations   67

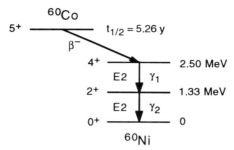

Fig. 5.3 Decay scheme of ^{60}Co.

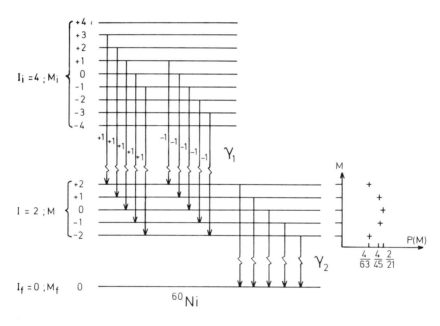

Fig. 5.4 Allowed transitions in ^{60}Ni when the z axis is chosen in the γ_1 emission direction (selection rule $m_1 = \pm 1$ for γ_1). The population probabilities $P(M)$ of the different sublevels of the intermediate state are shown on the right. The M_i and M sublevels are degenerate in energy but are shown separated for clarity.

rigorously. Again choosing the z axis as the emission direction of γ_1, we obtain the selection rule $m_1 = \pm 1$. For γ_2 of course, this limitation is not present, since this emission may occur at angles $\theta \neq 0$. In Fig. 5.4 the allowed transitions among sublevels are shown.

We will now use Eq. (5.4) to calculate the population probabilities $P(M)$ in the intermediate state. With $F_{21}(0) = F_{2-1}(0) = 1/2$ and using $3j$-symbol values from Rotenberg et al. (1959), we find

$$P(\pm 2) \propto \begin{pmatrix} 4 & 2 & 2 \\ \mp 3 & \pm 1 & \pm 2 \end{pmatrix}^2 + \begin{pmatrix} 4 & 2 & 2 \\ \mp 1 & \mp 1 & \pm 2 \end{pmatrix}^2 = \frac{1}{18} + \frac{1}{126} = \frac{4}{63}$$

$$P(\pm 1) \propto \begin{pmatrix} 4 & 2 & 2 \\ \mp 2 & \pm 1 & \pm 1 \end{pmatrix}^2 + \begin{pmatrix} 4 & 2 & 2 \\ 0 & \mp 1 & \pm 1 \end{pmatrix}^2 = \frac{4}{63} + \frac{8}{315} = \frac{4}{45}$$

$$P(0) \propto \begin{pmatrix} 4 & 2 & 2 \\ -1 & +1 & 0 \end{pmatrix}^2 + \begin{pmatrix} 4 & 2 & 2 \\ +1 & -1 & 0 \end{pmatrix}^2 = \frac{3}{63} + \frac{3}{63} = \frac{2}{21} \quad (5.9)$$

As expected, $P(M) = P(-M)$, but $P(M) \neq P(M')$ if $|M| \neq |M'|$. The $3j$-symbols for the transition from M to M_f are needed for the calculation of $W(\theta)$ using Eq. (5.6). These $3j$-symbols are all equal and are all

$$\begin{pmatrix} 2 & 2 & 0 \\ -M & m_2 & 0 \end{pmatrix}^2 = \frac{1}{5} \quad \text{for all } M, m_2 \text{ where } M = m_2$$

Substituting these values, one obtains

$$W(\theta) \propto 2 \times \frac{1}{5} P(\pm 2) F_{2\pm 2}(\theta) + 2 \times \frac{1}{5} P(\pm 1) F_{2\pm 1}(\theta) + \frac{1}{5} P(0) F_{20}(\theta)$$

(5.10)

Using the $P(M)$ values from Eq. (5.9) and omitting common factors

$$W(\theta) \propto 2 \times \frac{4}{63} \times \frac{1}{4}(1 - \cos^4 \theta) + 2 \times \frac{4}{45} \times \frac{1}{4}(1 - 3\cos^2 \theta + 4\cos^4 \theta)$$

$$+ \frac{2}{21} \times \frac{3}{2}(1 - \cos^2 \theta) \cos^2 \theta \quad (5.11)$$

Gathering terms and normalizing so the constant term is unity, we obtain for the γ-γ angular correlation of the ^{60}Ni cascade resulting from ^{60}Co β-decay,

$$W(\theta) = 1 + \frac{1}{8}\cos^2 \theta + \frac{1}{24}\cos^4 \theta \quad (5.12)$$

The naive theory considered above has a number of deficiencies:

(a) It is applicable only for pure transitions (only a single l). If more than one l value is important in the γ-ray transitions, the naive theory neglects interferences, because it uses transition probabilities, not the amplitudes.

(b) The choice of the z axis in the γ_1 emission direction imposes a restriction that makes it difficult to introduce perturbations (e.g. a magnetic field).
(c) Extranuclear perturbations introduce interferences in the transitions, even in the case of pure multipole transitions. Therefore transition amplitudes, not probabilities, must be computed properly in the theory.

5.1.2 General Theory

A few steps in the development of the general theory of γ–γ angular correlation are discussed in this section. This theory retains an arbitrary coordinate system rather than making the special choice of z in the γ_1 emission direction. Thus the emissions of γ_1 and γ_2 are treated on an equal footing. The geometrical arrangement and angles utilized in the general theory are illustrated in Fig. 5.5, and a schematic representation is shown in Fig. 5.6.

The transition amplitudes for the γ-emission from the initial to the

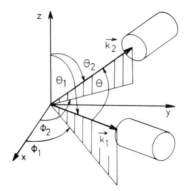

Fig. 5.5 Generalized coordinate system for the description of γ–γ angular correlation. k_1 and k_2 are the wave vectors of γ_1 and γ_2 respectively.

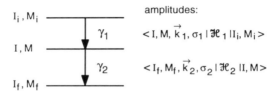

Fig. 5.6 Schematic representation of a γ–γ cascade. The matrix elements of the transitions are shown on the right. k_i and σ_i are the wave vector and polarization of γ_i ($i = 1, 2$).

intermediate state and for the intermediate to the final state depend on the matrix elements

$$\langle I, M, \mathbf{k}_1, \sigma_1 | \mathcal{H}_1 | I_i, M_i \rangle \quad \text{and} \quad \langle I_f, M_f, \mathbf{k}_2, \sigma_2 | \mathcal{H}_2 | I, M \rangle \quad (5.13)$$

where $\mathbf{k}_1$ is the wave vector and σ_1 the polarization of γ_1, and $\mathbf{k}_2$ and σ_2 the corresponding quantities for γ_2. $\mathcal{H}_1$ and $\mathcal{H}_2$ are the interaction Hamiltonian operators for the emission of γ_1 and γ_2.

Explicit calculation of the matrix elements in Eq. (5.13) is tedious. We discuss only the main features and leave out details.

Using the shorthand notation for these matrix elements,

$$\langle M | \mathcal{H}_1 | M_i \rangle \quad \text{and} \quad \langle M_f | \mathcal{H}_2 | M \rangle. \quad (5.14)$$

the angular correlation for a specific transition from M_i to M_f has the form

$$W(M_i \to M_f) = \left| \sum_M \langle M_f | \mathcal{H}_2 | M \rangle \langle M | \mathcal{H}_1 | M_i \rangle \right|^2 \quad (5.15)$$

Since the intermediate state M is not observed directly, the transition amplitudes, not the transition probabilities, must be summed over M. Thus interference terms not present in the naive theory appear in the rigorous general theory. Finally we must sum the transition probabilities over initial and final M values and the γ-ray polarizations. The probabilities are summed, not the amplitudes, because the initial and final M levels and polarizations can, at least in principle, be observed directly. One obtains for the angular correlation

$$W(\mathbf{k}_1, \mathbf{k}_2) = \sum_{M_i, M_f, \sigma_1, \sigma_2} \left| \sum_M \langle M_f | \mathcal{H}_2 | M \rangle \langle M | \mathcal{H}_1 | M_i \rangle \right|^2 \quad (5.16)$$

After calculating the matrix elements and summing (Frauenfelder and Steffen, 1965), one obtains

$$W(\mathbf{k}_1, \mathbf{k}_2) = W(\theta) = \sum_{\substack{k \text{ even}}}^{k_{\max}} A_k(1) A_k(2) P_k(\cos \theta) \quad (5.17)$$

Here k is a summation index that has the values

$$0 \leq k \leq \text{minimum of } (2I, l_1 + l_1', l_2 + l_2') \quad (5.18)$$

where I is the spin of the intermediate nuclear state and $l_{1,2}$ and $l'_{1,2}$ are the multipolarities of the transitions. The coefficient $A_k(1)$ depends only on the first transition and $A_k(2)$ depends only on the second. Their values have been tabulated (see e.g. Ferguson, 1965).

To this point we have considered only the possibility that unequal population of the $|I, M\rangle$ levels arises from detection of γ_1. However, this is not the only possibility (see Chapter 7). In the general case it is preferable to use a density matrix approach. The angular correlation is then written as

$$W(\mathbf{k}_1, \mathbf{k}_2) = W(\theta) = \sum_{\substack{k \text{ even}}}^{k_{max}} \rho_k(I) A_k(2) P_k(\cos \theta) \qquad (5.19)$$

with the statistical tensor

$$\rho_k(I) = \sqrt{(2k+1)(2I+1)} \sum_M (-)^{I-M} \begin{pmatrix} I & I & k \\ M & -M & 0 \end{pmatrix} P(M) \qquad (5.20)$$

where $P(M)$ are the probabilities introduced earlier. If $P(M)$ is determined by detection of the first γ-ray, then $\rho_k(I) = A_k(1)$.

5.2 Theory of Perturbed γ–γ Angular Correlations

We now consider a γ cascade for which the intermediate state has a certain lifetime τ_N (Fig. 5.7). If the nucleus is subject to a hyperfine interaction, the interaction could cause a substantial repopulation or phase change of the magnetic substates before the second γ-ray is emitted. If so then γ_2-rays will be emitted from sublevels with populations that exist at the time of the decay. The time evolution of a given substate $|M_a\rangle$ of the intermediate state under the influence of the time-development operator $\Lambda(t)$ is (index a refers

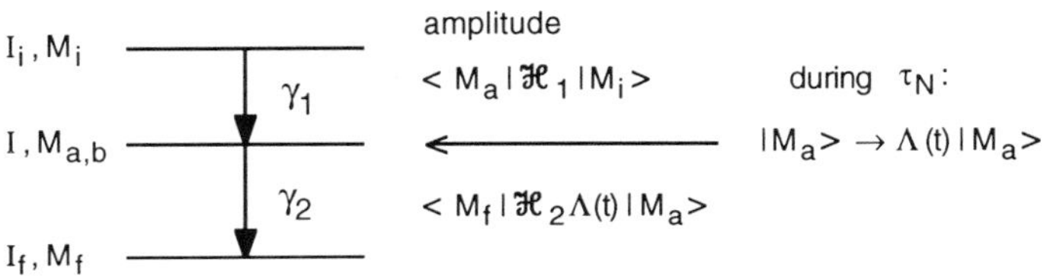

Fig. 5.7 The γ–γ cascade with mean lifetime τ_N in the intermediate state. The first γ-ray populates the state $|M_a\rangle$. During the lifetime of this intermediate state, a repopulation, or phase change, occurs.

72 Perturbed γ–γ Angular Correlation (PAC)

to the initially populated sublevel, and b to the sublevels that evolve from $|M_a\rangle$)

$$|M_a\rangle \to \Lambda(t)|M_b\rangle = \sum_{M_b}|M_b\rangle\langle M_b|\Lambda(t)|M_a\rangle \qquad (5.21)$$

In order to introduce the perturbation into the angular correlation, we begin with Eq. (5.16). We must change this equation to account for the change of the sublevel population with time between the two γ-rays, i.e. by replacing $|M\rangle$ by the perturbed state $\Lambda(t)|M\rangle$

$$W(\mathbf{k}_1, \mathbf{k}_2, t) = \sum_{M_i, M_f, \sigma_1, \sigma_2} \left|\sum_{M_a}\langle M_f|\mathcal{H}_2\Lambda(t)|M_a\rangle\langle M_a|\mathcal{H}_1|M_i\rangle\right|^2 \qquad (5.22)$$

With Eq. (5.21) for $\Lambda(t)|M\rangle$ and writing the square as a double sum with indices M and M'

$$W(\mathbf{k}_1, \mathbf{k}_2, t) = \sum_{\substack{M_i, M_f, \sigma_1, \sigma_2, \\ M_a, M'_a, M_b, M'_b}} \langle M_f|\mathcal{H}_2|M_b\rangle\langle M_b|\Lambda(t)|M_a\rangle\langle M_a|\mathcal{H}_1|M_i\rangle$$

$$\times \langle M_f|\mathcal{H}_2|M'_b\rangle^*\langle M'_b|\Lambda(t)|M'_a\rangle^*\langle M'_a|\mathcal{H}_1|M_i\rangle^* \qquad (5.23)$$

where M_a, M'_a, M_b, and M'_b are the M quantum numbers of the intermediate state.

As one can see, the influence of the perturbation in the intermediate state shows up only in the time-dependent factor $\langle M_a|\Lambda(t)|M_b\rangle\langle M'_a|\Lambda(t)|M'_b\rangle$. After a tedious calculation, one obtains the following rigorous form for the time-dependent γ–γ angular correlation,

$$W(\mathbf{k}_1, \mathbf{k}_2, t) = \sum_{k_1, k_2, N_1, N_2} A_{k_1}(1) A_{k_2}(2) G_{k_1 k_2}^{N_1 N_2}(t) \frac{1}{\sqrt{(2k_1+1)(2k_2+1)}}$$

$$\times Y_{k_1}^{N_1*}(\theta_1, \phi_1) Y_{k_2}^{N_2}(\theta_2, \phi_2) \qquad (5.24)$$

with the perturbation factor
$G_{k_1 k_2}^{N_1 N_2}(t)$

$$= \sum_{M_a, M_b} (-)^{2I + M_a + M_b} \sqrt{(2k_1+1)(2k_2+1)}$$

$$\times \begin{pmatrix} I & I & k_1 \\ M'_a & -M_a & N_1 \end{pmatrix} \begin{pmatrix} I & I & k_2 \\ M'_b & -M_b & N_2 \end{pmatrix} \langle M_b|\Lambda(t)|M_a\rangle\langle M'_b|\Lambda(t)|M'_a\rangle^*$$

$$(5.25)$$

The k_i are restricted to $k_i = 0, 2, \ldots,$ Min $(2I, l_i + l'_i)$ with $i = 1, 2$. The N_i are restricted to $|N_i| \leq k_i$.

As will be shown below, Eqs. (5.24) and (5.25) reduce to Eq. (5.17) in the limit of vanishing perturbation. When the perturbation of the intermediate state vanishes, then

$$\langle M_b | \Lambda(t) | M_a \rangle = \delta_{M_a, M_b}$$
$$\langle M'_b | \Lambda(t) | M'_a \rangle = \delta_{M'_a, M'_b} \quad (5.26)$$

where δ is the Kronecker delta symbol. This also implies $N_1 = N_2 =: N$. This gives a perturbation factor

$$G^{NN}_{k_1 k_2}(t) = \sum_{M_a} \sqrt{(2k_1 + 1)(2k_2 + 1)} \begin{pmatrix} I & I & k_1 \\ M'_a & -M_a & N \end{pmatrix} \begin{pmatrix} I & I & k_2 \\ M'_a & -M_a & N \end{pmatrix}$$
$$= \delta_{k_1, k_2} \quad (5.27)$$

so, for no interaction

$$G^{N_1 N_2}_{k_1 k_2}(t) = \delta_{k_1, k_2} \delta_{N_1, N_2} \quad (5.28)$$

Using the addition theorem for the spherical harmonics

$$W(k_1, k_2) = \sum_{k, N} A_k(1) A_k(2) \frac{1}{2k+1} Y^{N*}_k(\theta_1, \phi_1) Y^N_k(\theta_2, \phi_2)$$
$$= \sum_k A_k(1) A_k(2) P_k(\cos \theta) \quad (5.29)$$

where θ is the angle between k_1 and k_2. This result is just the unperturbed angular correlation given in Eq. (5.17).

5.3 Calculation of the Perturbation Factor for Special Cases

It is not difficult to calculate numerically the angular correlation from the general expressions given in Eqs. (5.24) and (5.25). However, for practical applications and for practical understanding, it is useful to derive analytic expressions for some common cases. A major simplification results if we restrict attention to *static, axially symmetric* interactions, and we make this restriction in the following discussion.

This condition is fulfilled of course for an external homogeneous magnetic field. Internal fields in magnetic materials also have axial symmetry, but the field direction can vary spatially. In this case, an average over magnetic field directions must be made in order to calculate the angular correlation. This case and non-axially-symmetric electric field gradients are excluded by our assumptions.

Choosing the z axis in the direction of the axial field, the matrix elements of the time-evolution operator are

$$\langle M_b|\Lambda(t)|M_a\rangle = \langle M_b|\exp\left(-\frac{i}{\hbar}\mathcal{H}t\right)|M_a\rangle$$

$$= \exp\left[-\frac{i}{\hbar}E(M)t\right]\delta_{M,M_a}\delta_{M,M_b} \quad (5.30)$$

where $M_a = M_b =: M$. A similar relation is valid for $\langle M'_b|\Lambda(t)|M'_a\rangle$. One sees clearly an advantage of the general formulation of the angular correlation theory that does not restrict the coordinate system as does the naive theory. Therefore we can utilize the symmetry of the hyperfine interaction to obtain the simple form of Eq. (5.30). One sees from Eq. (5.30) that, in this coordinate system, no repopulation of the M states occurs with time, just a phase change.

Substituting Eq. (5.30) into Eq. (5.25), we obtain the perturbation factor for static axial interactions,

$$G^{NN}_{k_1k_2}(t) = \sum_M \sqrt{(2k_1+1)(2k_2+1)} \begin{pmatrix} I & I & k_1 \\ M' & -M & N \end{pmatrix}\begin{pmatrix} I & I & k_2 \\ M' & -M & N \end{pmatrix}$$

$$\times \exp\left\{-\frac{i}{\hbar}[E(M) - E(M')]t\right\} \quad (5.31)$$

Only the energy differences between M sublevels appear in the perturbation factor, so a measurement of this factor allows one to determine these energy differences, and consequently the hyperfine interaction.

5.3.1 Magnetic Dipole Interaction

Eq. (3.3) gives the energy differences between the M levels due to a magnetic dipole interaction

$$E_{\text{magn}}(M) - E_{\text{magn}}(M') = -(M-M')g\mu_N B_z = N\hbar\omega_L \quad (5.32)$$

with $N = M - M'$. Substitution of Eq. (5.32) in Eq. (5.31) yields

$$G_{k_1k_2}^{NN}(t) = \sqrt{(2k_1 + 1)(2k_2 + 1)} \exp(-iN\omega_L t)$$

$$\times \sum_M \begin{pmatrix} I & I & k_1 \\ M' & -M & N \end{pmatrix} \begin{pmatrix} I & I & k_2 \\ M' & -M & N \end{pmatrix} \quad (5.33)$$

Using the orthogonality relations of the 3j-symbols, the sum over M in Eq. (5.33) gives $1/\sqrt{(2k_1 + 1)(2k_2 + 1)}\, \delta_{k_1,k_2}$, so we obtain for the magnetic dipole interaction

$$G_{kk}^{NN}(t) = \exp(-iN\omega_L t) \quad (5.34)$$

N and k are summation indices with the restrictions $|N| \leq k$, and k restricted to the values shown in Eq. (5.18). In particular, $|N| \leq 2I$ where I is the spin of the intermediate nuclear state. One sees from Eq. (5.34) that the angular correlation may, in general, precess at the fundamental frequency ω_L as well as higher harmonics $N\omega_L$.

A particularly simple expression is obtained for the angular correlation when the detector plane is perpendicular to the magnetic field $\mathbf{B}$. Substituting Eq. (5.34) in Eq. (5.24) gives for $\theta_1 = \theta_2 = 90°$ and with $\theta = \phi_1 - \phi_2$,

$$W_\perp(\theta, t, B_z) = \sum_{k \text{ even}}^{k_{max}} A_k(1) A_k(2) P_k[\cos(\theta - \omega_L t)] \quad (5.35)$$

It is sometimes more convenient to use cosines rather than Legendre polynomials, in which case the above equation becomes

$$W_\perp(\theta, t, B_z) = \sum_{k \text{ even}}^{k_{max}} b_k \cos[k(\theta - \omega_L t)] \quad (5.36)$$

For $k_{max} = 4$ and with the abbreviation $A_{kk} := A_k(1)A_k(2)$,

$$b_0 = 1 + \frac{1}{4}A_{22} + \frac{9}{64}A_{44}$$

$$b_2 = \frac{3}{4}A_{22} + \frac{5}{16}A_{44} \quad (5.37)$$

$$b_4 = \frac{35}{64}A_{44}$$

The perturbed angular correlation for $k_{max} = 2$ (e.g. only dipole transitions) and $\theta = 180°$ is

$$W_\perp(\theta = 180°, t, B_z) = b_0 + b_2 \cos(2\omega_L t) \quad (5.38)$$

Thus, the coincidence count rate shows a time modulation with twice the Larmor precession frequency.

5.3.2 Electric Quadrupole Interaction

We again restrict attention to the special case of static axially symmetric interactions. The energy separation between two M states for the electric quadrupole interaction (see Eq. (3.40)) is

$$E_Q(M) - E_Q(M') = 3|M^2 - M'^2|\hbar\omega_Q \tag{5.39}$$

Substitution into Eq. (5.31) yields for the perturbation factor

$$G^{NN}_{k_1 k_2}(t) = \sqrt{(2k_1 + 1)(2k_2 + 1)} \sum_M \begin{pmatrix} I & I & k_1 \\ M' & -M & N \end{pmatrix} \begin{pmatrix} I & I & k_2 \\ M' & -M & N \end{pmatrix}$$
$$\times \exp(-3i|M^2 - M'^2|\omega_Q t) \tag{5.40}$$

The perturbation factor in Eq. (5.40) can be rewritten more clearly as

$$G^{NN}_{k_1 k_2}(t) = \sum_n s^{k_1 k_2}_{nN} \cos(n\omega_Q^0 t) \tag{5.41}$$

where $\hbar\omega_Q^0$, the smallest nonvanishing energy difference between M levels (see Eq. (3.42)) is

$$\begin{aligned} \omega_Q^0 &= 3\omega_Q \quad \text{and} \quad n = |M^2 - M'^2| & \text{for } I \text{ integral} \\ \omega_Q^0 &= 6\omega_Q \quad \text{and} \quad n = (1/2)|M^2 - M'^2| & \text{for } I \text{ half-integral} \end{aligned} \tag{5.42}$$

For the parameters $s^{k_1 k_2}_{nN}$, one obtains

$$s^{k_1 k_2}_{nN} = \sqrt{(2k_1 + 1)(2k_2 + 1)} \sum_{M,M'} \begin{pmatrix} I & I & k_1 \\ M' & -M & N \end{pmatrix} \begin{pmatrix} I & I & k_2 \\ M' & -M & N \end{pmatrix}$$

$$\tag{5.43}$$

where the summation over M and M' should contain only values for which Eq. (5.42) is fulfilled.

The results obtained in Eq. (5.41) mean that the angular correlation rotates with frequencies ω_Q^0, $2\omega_Q^0$, ..., $n_{max}\omega_Q^0$ where each frequency appears with weight $s^{k_1 k_2}_{nN}$.

The polycrystalline sample (statistical distribution of electric field gradient orientations) is an important special case. In this case, the angular correlation depends only on the angle θ between k_1 and k_2. After some computation, one obtains the simple formula (see Frauenfelder and Steffen, 1965)

$$W(\theta, t) = \sum_{k \text{ even}}^{k_{\max}} A_{kk} G_{kk}(t) P_k(\cos \theta)$$

where   (5.44)

$$G_{kk}(t) = \sum_{n=0}^{n_{\max}} s_{kn} \cos(n\omega_Q^0 t)$$

In this special case (static axially symmetric electric field gradient in a polycrystalline sample), the perturbation factor $G_{kk}(t)$ does not depend on N; in addition, $k_1 = k_2 =: k$. The parameters s_{kn} are

$$s_{kn} = \sum_{M,M'} \begin{pmatrix} I & I & k \\ M' & -M & M-M' \end{pmatrix}^2 \qquad (5.45)$$

It might appear initially surprising that a time-dependent perturbation (precession) of the angular correlation remains observable even after the interaction is averaged over all electric field gradient orientations. The physical basis for such an effect is clarified graphically in Fig. 5.8. We decompose the precession around a given electric field gradient orientation into two components. One component is parallel to the symmetry axis of the angular correlation pattern. In this case, the angular distribution precesses around itself and shows no time dependence (hard core value). The other component is perpendicular to the symmetry axis and causes the time-dependent oscillation of the angular correlation. The oscillation frequency depends on the magnitude but not the direction of the electric field gradient.

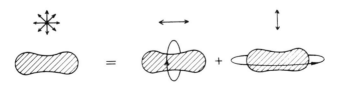

Fig. 5.8 Rotation of the angular correlation around different orientations of the electrical field gradient (arrows). Right: Decomposition into a time-constant fraction (rotation around the symmetry axis) and a time-varying fraction (rotation perpendicular to the symmetry axis of the angular correlation).

5.4 PAC Sources and Experimental Apparatus

5.4.1 PAC Sources

The most important requirement for a radioactive source to be useful for the PAC technique is the existence of an isomeric level with a lifetime between approximately 10 ns and several microseconds. The short time limit is set by the time resolution of experimental apparatus, and the long time limit is set by signal-to-noise considerations (the real coincidence to accidental coincidence count rate decreases as $1/\tau_N$). For electric field gradient investigations, the isomeric quadrupole moment should be large ($Q \geq 0.1$ b); for magnetic investigations, the magnetic dipole moment should be large ($\mu \geq \mu_N$).

In addition, of course, there must be a detectable γ-ray that populates the intermediate level and a second γ-ray emitted when it decays. The anisotropy of the γ–γ cascade must be large.

Finally the properties of the parent isotope, ease of production, physical and chemical properties, and in particular the lifetime (preferably several days to several weeks) are important.

The most commonly used PAC sources are ^{111}In, ^{181}Hf, and ^{100}Pd. The daughter isotopes, which are the actual probes measured in a PAC experiment, are ^{111}Cd, ^{181}Ta, and ^{100}Rh respectively. In the following we present the relevant properties of these three isotopes.

^{111}In–^{111}Cd

The isotope ^{111}In, which decays to ^{111}Cd by electron capture (EC) (Fig. 5.9), can be produced by the nuclear reactions ^{110}Cd(d, n)^{111}In, or ^{109}Ag-(α, 2n)^{111}In. The indium produced by either reaction can be separated chemically from the target material so the activity is essentially carrier-free.

The source is available commercially as indium chloride in dilute HCl solution. The indium may be introduced into a sample by diffusion, by implantation, or by incorporation during preparation from solution.

^{111}In is the most often used PAC source. Its importance in PAC is comparable to that of ^{57}Co in Mössbauer experimentation.

^{181}Hf–^{181}Ta

^{181}Hf can be produced easily by thermal neutron capture by ^{180}Hf. Its decay is shown in Fig. 5.10. Because of the high capture cross section ($\sigma = 14$ b)

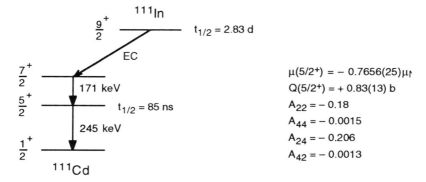

Fig. 5.9 Decay scheme of ^{111}In. The nuclear moments of the intermediate nuclear level (Lederer and Shirley, 1978; Vianden, 1983) and the anisotropy coefficients of the $\gamma-\gamma$ cascade (Raman and Kim, 1971; Steffen, 1956) are shown on the right.

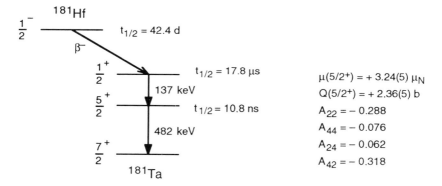

Fig. 5.10 Decay scheme of ^{181}Hf. The nuclear moments of the intermediate level (Lederer and Shirley, 1978; Butz and Lerf, 1983) and the anisotropy coefficients of the $\gamma-\gamma$ cascade (Ellis, 1973) are shown on the right.

one can obtain a large sample activity in a short time even with a small amount of hafnium. Although production is easy, separation of the radioactive ^{181}Hf from the starting material is difficult, since mass separation is required. To avoid the complications of mass separation and the necessity of using only samples with large hafnium concentration, one can increase the specific activity (ratio of radioactive to nonradioactive hafnium) by long irradiation times (approximately one month) at high neutron flux density. The radioactive hafnium is often introduced into samples by melting them together in ultra high vacuum ($p \leq 10^{-7}$ Pa). The good vacuum is required because of the large oxygen affinity of Hf.

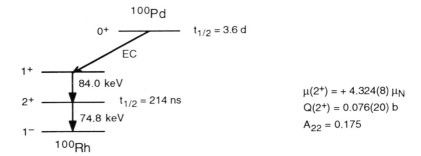

Fig. 5.11 Decay scheme of ^{100}Pd. The nuclear moments of the intermediate level (Lederer and Shirley, 1978; Vianden, 1983) and the anisotropy coefficients of the γ–γ cascade (Kocher, 1974) are shown on the right.

^{100}Pd–^{100}Rh

The isotope ^{100}Rh can be produced by a number of nuclear reactions. One example is ^{103}Rh(d, 5n)^{100}Pd with deuteron energy $E_d \geqslant 50$ MeV. Its decay is shown in Fig. 5.11. The chemical separation of ^{100}Pd from the rhodium target is possible in principle (Evans *et al.*, 1965) but difficult in practice because of the weak reactivity of these noble metals. Because of the separation difficulty, one often uses ion implantation to make samples, because the inactive rhodium atoms from the target can be separated from the ^{100}Pd by ion optics. The somewhat larger partial pressure of Pd compared to Rh facilitates ^{100}Pd evaporation in the ion source.

Compared to the two PAC sources discussed previously, ^{100}Pd has some disadvantages. First of all, the two γ-ray energies are almost the same and cannot be distinguished by NaI or BaF$_2$ detectors. Since the start and stop γ-rays are not distinguishable, one obtains time spectra in the multichannel analyzer that are symmetric around $t = 0$. Secondly, the integer spin ($I = 2$) in the intermediate state leads to a complicated angular correlation spectrum for non-axially-symmetric ($\eta \neq 0$) quadrupole interactions (10 different transition frequencies and a constant term).

5.4.2 *Experimental Apparatus*

In the above theoretical development, we have seen that the angular correlation between two γ-rays is time dependent in general. The time dependence is described by the perturbation factor (Eq. (5.25)) which contains all measurable information about the magnetic and quadrupole interactions in the intermediate state. Thus a measurement of this quantity provides complete information about these interactions. Therefore, in the following we describe the experimental apparatus with which the perturbation factor can be measured.

The time development of the angular correlation can be visualized qualitatively as a rotation of the radiation emission pattern. Detectors at fixed angles record coincident count rates that oscillate in time with frequencies that reflect the hyperfine interaction. The emission pattern sweeps past the detectors like a lighthouse beam.

Apparatus for time-differential perturbed angular correlation (TDPAC) is similar, in principle, to the angular correlation apparatus shown in Fig. 5.1. However, one does not measure the coincidence count rate as a function of angle but as a function of time between the emission of γ_1 and γ_2 at fixed detector angles (Fig. 5.12). Therefore, accurate measurement of this time difference is essential.

Two signals are derived from each detector. The anode signal is transformed to a digital timing pulse by a constant fraction discriminator (see Section 5.4.3). This time signal is delayed and sent to a coincidence circuit which allows it to pass if the γ-ray energy has the desired value. The energy of the γ-ray is determined by the detector dynode signal. That signal is amplified and sent to a single-channel analyzer which selects those signals with the desired energies. After the coincidence, the signal contains both time and energy information. By this fast/slow principle, energy selection (whether γ_1 or γ_2) and time measurement are performed separately.

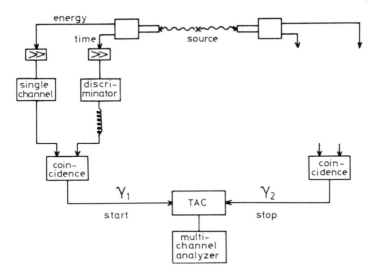

Fig. 5.12 Experimental apparatus for a time-differential perturbed angular correlation measurement. In the multichannel analyzer one obtains a count rate of form $N(\theta, t) = N_0 \exp(t/\tau_N) W(\theta, t)$, where τ_N is the lifetime of the intermediate state and $W(\theta, t)$ is the time-dependent angular correlation.

The signals from the coincidence circuits are used to start and stop a clock (time-to-amplitude converter (TAC), see Section 5.4.3). The TAC output is an analog signal whose height is converted to a digital value (analog-to-digital converter (ADC)) and stored in a multichannel analyzer.

The count rate of detector i is given by $N_i = \epsilon_i \Omega_i N$ where N is the source activity, ϵ_i the detector efficiency, and Ω_i the solid angle normalized to 4π. The coincident count rate N_{ij} between start detector i and stop detector j is then

$$N_{ij} = N_i \epsilon_j \Omega_j = \epsilon_i \epsilon_j \Omega_i \Omega_j N \tag{5.46}$$

This is the 'real' count rate arising from coincidences of γ_1 and γ_2 from the decay of the same nucleus. In addition there are 'random' coincidences arising from detection of a γ_1-ray and a γ_2-ray arising from different nuclei. This rate is

$$N_{ij}(\text{random}) = N_i N_j \tau = \epsilon_i \epsilon_j \Omega_i \Omega_j N^2 \tau \tag{5.47}$$

where τ is the total time-window recorded by the multichannel analyzer. The ratio of real to random coincidences is

$$\frac{N_{ij}}{N_{ij}(\text{random})} = \frac{1}{N\tau} \tag{5.48}$$

Since the optimal choice for τ is several times the nuclear lifetime τ_N, a typical value is $\tau = 1\,\mu s$. For $\tau = 1\,\mu s$, the source activity should not exceed 10^6 decays/s to obtain a real-to-random count ratio of more than 1.

The differential coincidence count rate at time t, where t is the elapsed time between the emission of γ_2 and γ_1, is

$$N_{ij}(\theta, t) = N_0 \exp(-t/\tau_N) W(\theta, t) + B \tag{5.49}$$

B is the time-independent background random coincident count rate; $W(\theta, t)$ represents the time-dependent angular correlation (θ is the angle between the two detectors). Examples of such spectra, $N(90°, t)$ and $N(180°, t)$, are shown later in Fig. 5.18 for the case of ^{111}Cd in Cd metal.

Usually, one uses four detectors in a plane at 90° angular separations (Fig. 5.13). Each detector can then be used as both a start and stop detector. These combinations lead to a total of 12 possible coincidence spectra.

By combining four spectra from which the background B has been subtracted, one can form the following

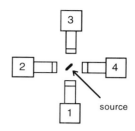

Fig. 5.13 Geometrical arrangement of the detectors for a four-detector PAC apparatus.

$$R(t) = \frac{2}{3}\left[\sqrt{\frac{N_{13}(180°, t)N_{24}(180°, t)}{N_{14}(90°, t)N_{23}(90°, t)}} - 1\right] \quad (5.50)$$

For axially symmetric, randomly oriented electric field gradients, the angular correlation is given by (Eq. (5.44))

$$W(\theta, t) = 1 + A_{22}G_{22}(t)P_2(\cos\theta) + \ldots \quad (5.51)$$

where the terms with $k > 2$ are often neglected. With this very simple angular correlation, the count-rate ratio $R(t)$ is given for $|A_{22}| \ll 1$ by

$$R(t) \approx A_{22}G_{22}(t) \quad (5.52)$$

Thus, we now have a relation that permits the determination of the desired quantity, the perturbation factor, from the experimentally measurable quantity $R(t)$. The detector efficiencies and solid angles do not appear in $R(t)$, since they appear in both, the numerator and denominator of Eq. (5.50) and consequently cancel. This is an important feature and makes the four-detector arrangement very convenient for PAC applications. An example of this count-rate ratio $R(t)$ is shown in the lower part of Fig. 5.18.

5.4.3 Electronic Apparatus for Time Measurement

Leading-edge discriminator

The arrival time of a γ-ray in a detector can be measured by the electronics only to a precision Δt. The time of the output signal from a leading-edge detector is determined by the time at which the input voltage crosses a previously set threshold (see Fig. 5.14). The discriminator level must be set at a high enough value that electronic noise does not cause excessive triggering. The nonzero level causes the output signal to be delayed relative

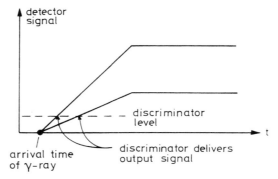

Fig. 5.14 Origin of the time jitter for different input signal pulse heights as detected by a leading-edge discriminator.

to the 'real' arrival time. This delay depends on the pulse height (see Fig. 5.14). Thus, the leading-edge principle leads to an unavoidable time jitter.

Constant-fraction discriminator

To eliminate the pulse height dependence that causes time jitter for leading-edge discrimination, one uses a constant-fraction discriminator. The principle of the constant-fraction discriminator is shown in Fig. 5.15. The input signal is split; one part is delayed, the other part inverted, attenuated, and then added to the first part. The shape of the resulting signal is shown in the lower part of Fig. 5.15. One sees that the resulting signal crosses zero at a time independent of the pulse height. A zero-crossing discriminator outputs a pulse at the time of the zero crossing.

Time-to-amplitude converter (TAC)

The time-to-amplitude converter (TAC) is an electronic stopwatch. It converts the time difference between two digital input signals into an output signal whose pulse height is proportional to that time difference. The analog output signal can then be digitized by an analog-to-digital converter and stored in a multichannel analyzer. The principle is shown in Fig. 5.16.

A time-to-digital converter (TDC) is another possibility for measuring the time difference between two input pulses. In a TDC, a counter is started by the first signal and stopped by the second. The count is proportional to the time difference. This method has the advantage that the digital number can be transferred directly to a computer.

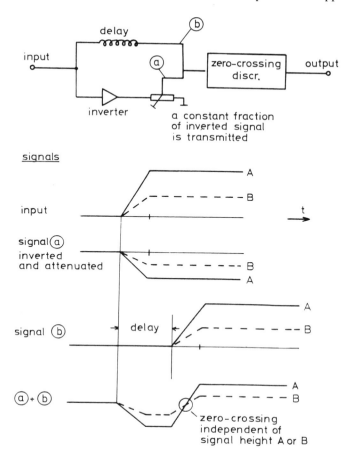

Fig. 5.15 Top: Block diagram of a constant-fraction discriminator. Bottom: Signal shape at the points a and b and after combination of the two signals a + b.

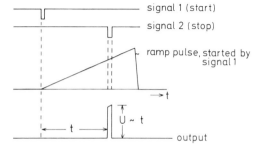

Fig. 5.16 Principles of operation for a time-to-amplitude converter (TAC): the first signal starts a ramp pulse; at the time of the second pulse, the voltage of the ramp is measured.

5.5 Electric Field Gradients in Non-Cubic Metals

A probe nucleus at a substitutional lattice site in a non-cubic metal experiences an electric field gradient because of the non-cubic arrangement of nearby atomic cores and electrons. We will discuss the electric field gradient for ^{111}In probes in Cd metal (Fig. 5.17).

Cadmium has a hexagonal crystal structure ($c/a = 1.89$ at RT) which produces an axially symmetric ($\eta = 0$) electric field gradient. For a polycrystalline sample (many small crystallites) the electric field gradients are randomly oriented. After the decay of ^{111}In to ^{111}Cd, the electronic shell of the probe relaxes within approximately 10^{-13} s to that of the host. Therefore one observes the electric field gradient at a cadmium nucleus in the cadmium metal. PAC spectra for ^{111}Cd in cadmium metal are shown in Fig. 5.18. The coincidence count rates $N(\theta, t)$ for the two detector geometries, $\theta = 90°$ and $\theta = 180°$, are shown in the upper part of the figure. The count-rate ratio $R(t)$ (see Eqs. (5.50) and (5.52)) is shown in the lower part. The experimental spectrum $R(t)$ was fitted with the function

$$A_{22}G_{22}(t) = A_{22} \sum_{n=0}^{3} s_{2n} \cos(n\omega_Q^0 t) \tag{5.53}$$

Since the nuclear spin $I = 5/2$, only three frequencies appear and have frequency ratio $1:2:3$ (see Section 3.2). For $I = 5/2$, the parameters s_{2n} have the following values:

$$s_{20} = 0.20 \quad s_{21} = 0.37 \quad s_{22} = 0.29 \quad s_{23} = 0.14 \tag{5.54}$$

The quadrupole coupling constant obtained in this experiment (Witthuhn and Engel, 1983) is

$$\nu_Q = \frac{10}{3\pi} \omega_Q^0 = 124.7(5) \text{ MHz} \tag{5.55}$$

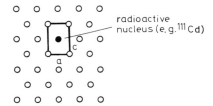

Fig. 5.17 Environment of a radioactive nucleus in a hexagonal crystal lattice (e.g. Cd metal: $c/a = 1.89$).

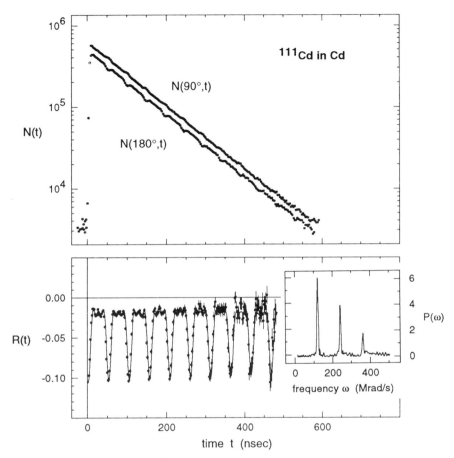

Fig. 5.18 PAC spectrum for ^{111}Cd in cadmium metal. The coincidence count rates $N(\theta, t)$ for $\theta = 90°$ and $\theta = 180°$ are shown in the upper part of the figure. The count-rate ratio $R(t)$ is shown in the lower part. The solid line is fitted to the theoretical function, Eq. (5.53). The inset shows a Fourier analysis of the spectrum $R(t)$, where the three frequencies can be seen clearly.

Using the quadrupole moment of the ^{111}Cd intermediate state one obtains for the electric field gradient magnitude (see Eq. (3.43))

$$V_{zz} = 6.21(75) \times 10^{21} \text{ V m}^{-2} \quad (\eta = 0) \tag{5.56}$$

The values were obtained at room temperature ($T = 295$ K).

Calculation of the electric field gradient is difficult, particularly in metals. One must calculate the contributions to the electric field gradient from neighboring ionic cores, the electrons of the probe core, and the conduction electrons. The most difficult problem is to compute the spatial distribution of conduction electrons, since it requires extremely accurate band-structure calculations. Self-consistent molecular cluster calculation, in which the solid is approximated by a small atomic cluster (up to 30 atoms), is an alternative approach. Using this model, Lindgren (1986) obtained for Cd in cadmium a theoretical electric field gradient $V_{zz}(\text{theor}) \approx 5 \times 10^{21}$ V m^{-2} ($T = 0$ K) in reasonable agreement with experiment.

One often can do no better than use a phenomenological model. The electric field gradient is assumed to be a sum of two parts, one from the neighboring ion cores (lattice), and one from the electrons (el)

$$V_{zz} = (1 - \gamma_\infty)V_{zz}(\text{lattice}) + (1 - R)V_{zz}(\text{el}) \tag{5.57}$$

An amplification factor $(1 - \gamma_\infty)$ and $(1 - R)$ is assumed for the lattice and electronic contribution respectively. The factor γ_∞, called the Sternheimer antishielding factor, accounts for the probe atom inner-shell polarization. The index ∞ means that the field gradient sources are completely outside the probe atomic shell. This assumption is not rigorously valid. Commonly one obtains $(1 - \gamma_\infty)$ values of order 10 to 100 (Feiock and Johnson, 1969), i.e. the amplification is very large and thus determines the magnitude of the electric field gradient.

An analogous factor $(1 - R)$ is introduced to account for amplification of the electronic part of the electric field gradient.

V_{zz}(lattice)

The electric field gradient arising from the ion cores can easily be calculated if these cores are assumed to be point charges. Contributions of all lattice atoms, except for the probe atom itself, must be summed. This yields

$$V_{\alpha\beta} = \frac{Ze}{4\pi\epsilon_0} \sum_{i \neq 0} \frac{1}{r_i^5} \begin{pmatrix} 3x_i^2 - r_i^2 & 3x_iy_i & 3x_iz_i \\ 3x_iy_i & 3y_i^2 - r_i^2 & 3y_iz_i \\ 3x_iz_i & 3y_iz_i & 3z_i^2 - r_i^2 \end{pmatrix} \tag{5.58}$$

where r_i is the vector from the origin to the point charge i.

Temperature dependence of the electric field gradient

A rather large temperature dependence has been observed for many systems. This dependence is too large to be accounted for by thermal expansion. Christiansen *et al.* (1976) found that the electric field gradient in a number of non-cubic metals follows an empirical $T^{3/2}$ temperature dependence

$$V_{zz}(T) = V_{zz}(0)(1 - BT^{3/2}) \tag{5.59}$$

Although the $T^{3/2}$ temperature dependence is not quantitatively understood, it has been attributed to lattice vibrations. In that model, the temperature dependence is described by

$$V_{zz}(T) = V_{zz}(0)[1 - \alpha \langle u^2 \rangle (T)] \tag{5.60}$$

with $\langle u^2 \rangle (T)$ being the average square atomic displacement.

Within this model, the electric field gradient $T^{3/2}$ dependence arises from a $T^{3/2}$ dependence of the mean square atomic displacement $\langle u^2 \rangle$. In the Debye model, $\langle u^2 \rangle$ is proportional to T^2 at low temperature and to T at high temperature (see Section 4.7), which crudely approximates a $T^{3/2}$ dependence.

5.6 Atomic Defects in Metals

In cubic metals, the electric field gradient at a lattice site is zero. However, if there are vacancies in the metal, for example if the material is damaged by irradiation, then PAC probes near such defects experience an electric field gradient. Different defect configurations (e.g. single vacancies, divacancies, interstitials, etc.) produce different electric field gradients, and therefore can be distinguished by PAC measurement.

As an example, we discuss the so-called neutrino-recoil experiment (Metzner et al., 1984) which provides particularly clear information about the production and annealing of defects.

The radioactive probe ^{111}Sn is produced by the nuclear reaction ^{93}Nb(^{22}Ne, 4n)^{111}Sn on a thin niobium foil and implanted into a copper backing by the recoil from the nuclear reaction. The copper is annealed so that all ^{111}Sn atoms are at substitutional lattice sites in defect-free surroundings. Then the sample is cooled to 4 K except for periods of annealing as discussed below. The perfect lattice situation is shown on the left side of Fig. 5.19.

The decay of ^{111}Sn proceeds by the following sequence, with the halflives as indicated:

$$^{111}\text{Sn} \xrightarrow{35 \text{ min}} {}^{111}\text{In} \xrightarrow{2.8 \text{ d}} {}^{111}\text{Cd}$$

Some 61% of the ^{111}Sn decays to the ^{111}In ground state by electron capture

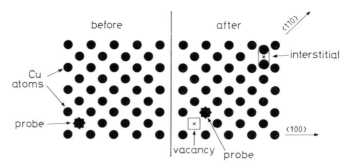

Fig. 5.19 Microscopic lattice arrangement before and after the neutrino decay of ^{111}Sn. Adapted from Metzner et al. (1987).

(EC) and emits a monoenergetic neutrino. Momentum conservation requires that the indium atom experience a recoil given by

$$E_{\text{recoil}} = \frac{Q^2}{2Mc^2} \qquad (5.61)$$

where $Q = 2436$ keV is the Q-value for K-electron capture, and M is the mass of ^{111}In. Substituting these values into Eq. (5.61) gives $E_{\text{recoil}} = 29$ eV.

The threshold for displacement of copper atoms in a copper lattice is known and is 19 eV. The recoil energy of 29 eV therefore can displace one copper atom and create a Frenkel pair (a vacancy and interstitial); however, the recoil energy is insufficient for two such displacements. The β^+-decay (39%) that competes with the EC process transfers a maximum of 17 eV and therefore cannot cause a copper-atom displacement.

Thus, in the ^{111}Sn decay, one has the fortunate situation that either just one Frenkel pair is created or none. In Fig. 5.19, we picture a displacement as a collision cascade along the closest-packed direction ($\langle 110 \rangle$), consistent with the result of the PAC experiment. In such a displacement, a vacancy is left at the original ^{111}Sn site, which is now in a first-neighbor position to a ^{111}In, and an interstitial is created some distance away. It is known that interstitials in copper exist in a dumbbell configuration with another atom (see the right side of Fig. 5.19).

After the ^{111}Sn decay, the resulting ^{111}In is left either with a first-neighbor vacancy or in a perfect, unperturbed lattice site.

Since the recoil energy resulting from the ^{111}In to ^{111}Cd decay is too small to cause further displacements, in the PAC experiment one observes an unperturbed fraction due to indium atoms in perfect lattice sites and a fraction with a quadrupole interaction due to a first-neighbor vacancy. The

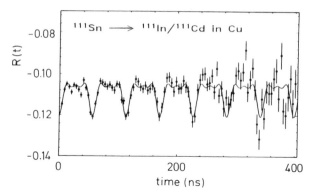

Fig. 5.20 PAC spectrum of ^{111}In after the neutrino decay of ^{111}Sn in copper. Adapted from Metzner *et al.* (1987).

experimental spectrum is shown in Fig. 5.20. One clearly sees an undamped oscillation with a constant offset.

This experiment creates a clear starting point for annealing experiments. The fraction of ^{111}In probes having a near-neighbor vacancy is shown in Fig. 5.21 as a function of isochronal annealing temperature (10 minute anneal).

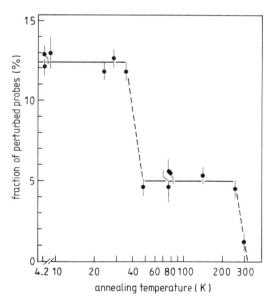

Fig. 5.21 Fraction of ^{111}In probes with a nearest-neighbor vacancy as a function of annealing temperature. The copper sample was annealed for 10 minutes at the indicated temperature, and then cooled again to the measuring temperature of 4 K. Adapted from Metzner *et al.* (1987).

One recognizes two steps. The first occurs between 30 and 40 K, the so-called stage-I region, where interstitials become mobile and recombine with vacancies, thereby reducing the number of vacancies. However, it is clear that not all vacancies disappear. This must arise because the number of interstitials created in the ^{111}Sn decay is insufficient to fill all vacancies, possibly because of competing processes (interstitial trapping at surfaces, grain boundaries or other defects). The second annealing step occurs near 250 K, the so-called stage-III region, where vacancies become mobile. At this temperature, complete annealing occurs as expected. If annealed at higher temperatures, one observes a fully unperturbed PAC spectrum.

5.7 Adsorbate Sites on Surfaces

Radioactive PAC probes at substitutional sites in the interior of a cubic crystal see no electric field gradients because of the crystal symmetry. This situation changes, however, if the probes are at or near a surface, where the environment is no longer cubic. PAC studies of surfaces are becoming increasingly important (Schatz et al., 1990).

As an example of such investigations we will describe the adsorption of ^{111}In atoms on silver surfaces (Fink et al., 1990). These surfaces must be monocrystalline and are prepared in ultrahigh vacuum. The quality, purity, and crystallinity are usually surveyed by Auger electron spectroscopy (AES) and low-energy electron diffraction/spectroscopy (LEED). For the case discussed here, stepped (111) Ag surfaces were prepared. The steps along a $\langle 001 \rangle$ direction were created by appropriate cutting of the original single crystal. Approximately 10^{11} ^{111}In atoms were deposited on this surface at 77 K (for comparison, there are approximately 10^{15} silver atoms/cm^2 of surface).

Immediately after deposition, the probes experience a relatively weak electric field gradient. The field gradients progress through different values with increasing annealing temperature, indicating that the ^{111}In probes move to successively different sites. This can be seen clearly in Fig. 5.22. The fractions of the distinguishable sites found in this experiment are designated as f_{Ad}, f_{S1}, f_{S2}, and f_0. The electric field gradients experienced by the fractions f_{Ad} and f_0 are axially symmetric with axis perpendicular to the surface. The other electric field gradients for f_{S1} and f_{S2} are non-axial with one principal axis along the step direction (see axis orientations on the right side of Fig. 5.23).

The probe fractions are shown as a function of annealing temperature on the left side of Fig. 5.23 (measuring temperature always 77 K). One sees clearly that a succession of different sites are populated with increasing annealing temperature. The temperature dependence, along with the

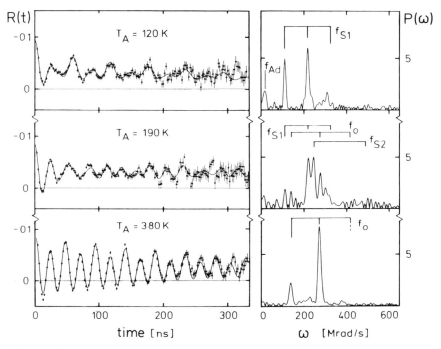

Fig. 5.22 PAC spectra along with their Fourier transforms for ^{111}In on stepped Ag(100) for three different annealing temperatures (annealing time 15 minutes, measuring temperature 77 K). The solid lines were fitted to the experimental points by Eq. (5.41).

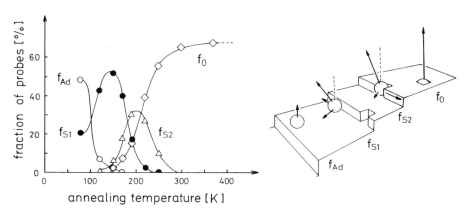

Fig. 5.23 Left side: Fraction of indium probes with different electric field gradients on stepped Ag(100) as a function of annealing temperature. Right side: Surface site model for indium probes suggested by the experimental parameters.

measured electric field gradient parameters, leads to the following model for the adsorption sites (see right side of Fig. 5.23). On deposition at low temperature, the ^{111}In probes are adatoms (f_{Ad}) on the flat surface. On annealing, they migrate to steps where they first occupy adatomic step sites (f_{S1}). After higher-temperature annealing, the probes are incorporated into the steps (f_{S2}). With still higher-temperature annealing, the probes diffuse, via vacancy diffusion, into the first monolayer of the surface (f_0).

This model for the surface sites was further supported by calculations of the electric field gradients (Lindgren, 1990). In addition, the site assignments were confirmed by low-energy ion scattering (Breman and Boerma, 1991).

5.8 Internal Magnetic Fields in Ferromagnetic Materials

Here we introduce the perturbed angular distribution (PAD) method, a variant of the PAC technique. In PAD, alignment of the nuclear spin in the isomeric state is not produced by observing γ_1 but rather by the angular momentum transfer in a nuclear reaction (see Fig. 5.24). The time at which the isomeric state is formed can be determined either by detecting a particle accompanying this nuclear reaction or by using a pulsed accelerator beam.

An example of PAD is the measurement of local magnetic fields in ferromagnetic materials. The B field at the nuclear site can be obtained directly from the precession frequency of the γ_2-ray angular distribution if the g-factor of the probe is known.

As examples of such experiments the $R(t)$ spectra of ^{67}Ge probes ($I = 9/2$

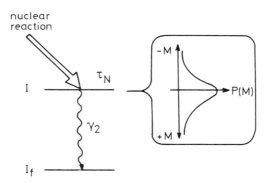

Fig. 5.24 Variant of the PAC method. The angular momentum transferred by a beam is perpendicular to the beam axis. If the z axis is taken as the beam direction, the substates with low M values are preferentially populated. The resulting anisotropic γ_2 angular distribution can precess under the influence of the hyperfine interaction.

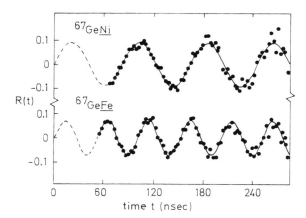

Fig. 5.25 The $R(t)$ spectra for ^{67}Ge probes in nickel and iron at 300 K, showing Larmor precession. Adapted from Raghavan et al. (1978).

isomer with lifetime $\tau_N = 101$ ns) in nickel and iron are shown in Fig. 5.25. The probes are produced by the reaction ^{54}Fe(^{16}O, 2pn)^{67}Ge with a pulsed ^{16}O beam (energy 53 MeV) and recoil-implanted into nickel or iron. The internal B fields in magnetic materials can be measured with high precision using this method.

The local B fields B_{hf} (hf denotes hyperfine) in iron measured by different probe nuclei are shown in Fig. 5.26. The magnetic field at the nucleus is determined by the spin polarization of the s electrons at the nuclear position. The dipole contributions from magnetic moments near the nucleus are much smaller and may be neglected. Fig. 5.26 also shows a theoretical calculation (Kanamori et al., 1981) of the internal B fields (solid line) which agrees well with experiment.

5.9 Integral Perturbed Angular Correlation (IPAC) and Transient Magnetic Fields in Ferromagnets

The integral PAC method (IPAC) is used if the experimental time resolution Δt is larger than or comparable to the lifetime τ_N of the intermediate state, i.e. if

$$\Delta t \geq \tau_N \tag{5.62}$$

In practice, nuclear decays with lifetime smaller than 1 ns can be measured only by IPAC. In this case the differential perturbation factor $G_{k_1 k_2}^{N_1 N_2}(t)$ must be integrated over time with a weighting function $\exp(-t/\tau_N)$. We limit

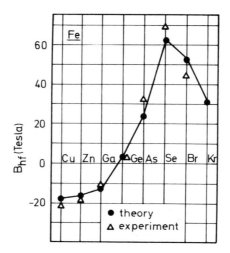

Fig. 5.26 Local B field, B_{hf}, at different probe nuclei in iron. Experimental values (triangles) are compared to theoretical values (closed circles, solid line) (from Kanamori et al., 1981).

ourselves to the measurement of B fields. For a magnetic interaction, one obtains for perpendicular geometry (see Eq. (5.36))

$$W_\perp(\theta, t, B_z) = \sum_k b_k \cos[k(\theta - \omega_L t)] \quad (5.63)$$

The integral γ–γ angular distribution is then

$$\overline{W_\perp(\theta, B_z)} = \frac{1}{\tau_N} \int_0^\infty \exp(-t/\tau_N) W_\perp(\theta, t, B_z)\, dt \quad (5.64)$$

Substituting Eq. (5.63) into Eq. (5.64) yields

$$\overline{W_\perp(\theta, B_z)} = \frac{1}{\tau_N} \sum_k b_k \int_0^\infty \exp(-t/\tau_N) \cos[k(\theta - \omega_L t)]\, dt$$

$$= \sum_k b_k \frac{\cos[k(\theta - \Delta\theta)]}{\sqrt{1 + (k\tau_N \omega_L)^2}} \quad (5.65)$$

with

$$\tan(k\Delta\theta) = k\tau_N \omega_L \quad (5.66)$$

For τ_N very small compared to $2\pi/\omega_L$, one obtains $k\tau_N\omega_L \ll 1$, yielding

$$\overline{W_\perp(\theta, B_z)} = \sum_k b_k \cos[k(\theta - \tau_N\omega_L)] \tag{5.67}$$

This result means that the angular correlation looks the same as the original correlation but that it is shifted by the angle $\Delta\theta = \tau_N\omega_L$. A schematic illustration of the apparatus and physical principles of this method is shown in Fig. 5.27 along with an experimental result for ^{192}Pt in iron. If the g-factor and lifetime τ_N are known, the rotation angle

$$\Delta\theta = \tau_N\omega_L = -\tau_N g \frac{\mu_N}{\hbar} B_z \tag{5.68}$$

can be used to determine the B field at the nucleus. This method is particularly well suited to measurement of very large internal magnetic fields. For such experiments, it is necessary that the internal fields be aligned perpendicular to the detector plane. This is achieved by applying a small external magnetic field in this direction.

Measurement of internal B fields by IPAC presents no particular problem. However, one often finds that the B fields measured by an IPAD experiment, for which the isomeric state is populated by an in-beam nuclear reaction, are different from those measured in an analogous γ–γ IPAC measurement on the same nuclear state. The deviation arises from different (transient) magnetic fields which act on the probe nucleus during its recoil motion after the nuclear reaction. The rotation angle of the γ-ray angular distribution after the in-beam reaction can be expressed approximately as a

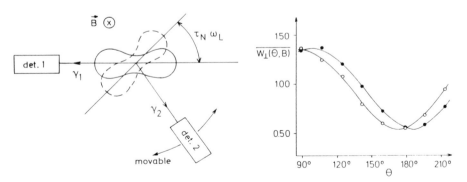

Fig. 5.27 Left: Schematic illustration of apparatus and physical principles of the IPAC method. Right: Experimental result for the 604–317 keV cascade of ^{192}Pt in iron for two opposite directions of the external B field (from Katayama et al., 1975).

sum of the rotation in the static field B_0 and the rotation in the transient field B_{trans}

$$\Delta\theta = \Delta\theta(B_0) + \Delta\theta(B_{\text{trans}}) = -g\frac{\mu_N}{\hbar}(B_0\tau_N + B_{\text{trans}}\tau_1) \quad (5.69)$$

where B_{trans} is the effective transient B field, and τ_1 is the effective interaction time. In general, the effective values arise from complicated averaging over the lifetime of the nuclear level and the slowing-down time in the ferromagnet. A simple relation is obtained if the lifetime of the nuclear level is so short that one can neglect slowing down during the lifetime. Another simple case is an experiment on a foil that is thin enough that the nuclear velocity (and consequently the transient field) can be considered constant in the foil. The interaction time with the transient field is determined by the nuclear level lifetime in the first case and by the flight time through the ferromagnetic foil in the second. We will discuss an example of the second case in more detail (Speidel, 1985).

The radioactive probe ^{16}O in the 3^- state ($E^* = 6.13$ MeV, $\tau_N = 26.6$ ps) was produced by the nuclear reaction ^{16}O(α, α'). The probe velocity is the recoil velocity of the nucleus after being struck by the α-particle. Higher velocities can be obtained using the inverse reaction ^{4}He(^{16}O, ^{16}O*) where ^{16}O is the beam and ^{4}He the target. The probes were stopped in (the nonmagnetic metal) silver after passing through a ferromagnetic foil.

In this experiment, the only significant angular correlation rotation occurs during the passage through the foil (thickness approximately 1 μm). Using Eq. (5.69) one can calculate the transient B field from the measured rotation angle. The effective interaction time τ_1 is the time the ^{16}O is within the foil (≤ 1 ps).

Results for different probe velocities, v/v_0 ($v_0 = c/137$), are shown in Fig. 5.28. The transient field experienced by the probe passing through iron is approximately 400 T and independent of probe velocity. In Gd, the transient field is also approximately 400 T at low velocity but increases to over 1000 T at higher velocity. The transient B fields are one or more orders of magnitude larger than typical static internal fields in Fe and Gd (compare with Fig. 5.26).

Transient B fields are not fully understood. We discuss a model (Dybdal et al., 1979; Speidel, 1985) that provides a qualitative understanding, at least for light ions. In this model it is assumed that the transient field is due to s electrons remaining bound to the nucleus. From other measurements (Dybdal et al., 1979) one knows that for velocities between $2v_0$ and $5v_0$ the probability of a single K-shell hole for ^{16}O in iron is approximately 27%. The remaining 1s electron causes a hyperfine field of 8550 T. If one assumes that the spin of this electron changes rapidly during passage through the

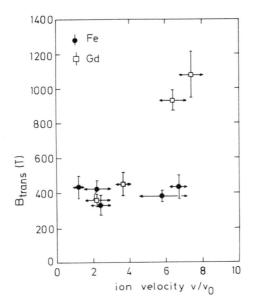

Fig. 5.28 Transient B field for ^{16}O nuclei during passage through Fe and Gd foils as a function of ^{16}O velocity (from Speidel, 1985). Velocity is shown in units of $v_0 = c/137$.

ferromagnetic foil but that on average a polarization parallel to the macroscopic magnetization is established, then the transient field is given by

$$B_{\text{trans}} = qpB_{1s} \tag{5.70}$$

where q is the probability of having just one 1s electron, p is its polarization, and B_{1s} the hyperfine field produced by a 1s electron. With $q = 0.27$ and $B_{1s} = 8550\,\text{T}$, one needs a polarization $p = 17\%$ to account for the experimentally-observed transient B field for ^{16}O in Fe. The mechanism of polarization transfer from the ferromagnet to the 1s electron is not specified in this model. The probability of a K-shell hole is not known for a Gd host, so similar calculations cannot be made.

6
Nuclear Magnetic Resonance (NMR)

6.1 Principles

Nuclear magnetic resonance (NMR) was invented in 1945 with major contributions from F. Bloch and E. M. Purcell (Bloch *et al.*, 1946; Purcell *et al.*, 1946).

Nuclear magnetic resonance is the oldest nuclear method used in solid state physics. It is also used extensively in chemistry, biology, medicine, and other technological fields. In medicine NMR tomography is becoming a common technique for imaging portions of the human body and is a totally non-invasive method; local NMR parameters are used to provide contrast images.

The basic idea of NMR is that radiofrequency signals can be used to measure resonant properties of nuclei in magnetic fields. In an external magnetic field the magnetic dipole moments of nuclei are partially aligned, and the magnetization can be changed by irradiating with a radiofrequency electromagnetic field. Nuclear magnetization and resonance absorption are important concepts and will be discussed.

Nuclear magnetization.

In a static magnetic field $\boldsymbol{B}_0$, the nuclear state with spin I splits into $(2I + 1)$ sublevels (see Section 3.1). In the absence of other fields, the previously degenerate M states are shifted by the energy $E_{\text{magn}}(M) = M\hbar\omega_L$ (where ω_L is the Larmor frequency).

In thermal equilibrium, the substates are occupied with probability given by the Boltzmann distribution

$$P(M) \propto \exp\left[-E_{\text{magn}}(M)/k_B T\right] \qquad (6.1)$$

Owing to the unequal population of the M states, one has a polarization of the nuclear spin

$$\langle I_z \rangle = \frac{\sum_{M=-I}^{I} \hbar M \exp[-E_{\text{magn}}(M)/k_B T]}{\sum_{M=-I}^{I} \exp[-E_{\text{magn}}(M)/k_B T]} \tag{6.2}$$

Except at very low temperature, $|E_{\text{magn}}(M)| \ll k_B T$, the exponential function can be expanded,

$$\langle I_z \rangle = \frac{\sum_{M=-I}^{I} \hbar M (1 + \gamma \hbar M B_0 / k_B T)}{\sum_{M=-I}^{I} (1 + \gamma \hbar M B_0 / k_B T)} \tag{6.3}$$

where $E_{\text{magn}}(M) = -\gamma \hbar M B_0$ (γ is the gyromagnetic ratio) was used. The sum $\sum M$ from $M = -I$ to $M = I$ is zero, so the first term in the numerator and the second term in the denominator disappear. This gives

$$\langle I_z \rangle = \frac{\gamma \hbar^2 B_0}{k_B T} \frac{\sum M^2}{2I + 1} = \frac{\gamma \hbar^2 I(I+1)}{3 k_B T} B_0 \tag{6.4}$$

The magnitude of the nuclear spin polarization $\langle I_z \rangle$ is very small at room temperature ($T = 300$ K), e.g. for $B_0 = 1$ T, $I = 1$, and $\gamma = \mu_N / \hbar$, one obtains

$$\frac{\langle I_z \rangle}{\hbar} = \frac{2 \mu_N B_0}{3 k_B T} \approx 10^{-6} \tag{6.5}$$

Some degree of nuclear polarization is essential for NMR, and although the polarization is small, it is entirely sufficient for observation of NMR signals.

The nuclear spin polarization $\langle I_z \rangle$ leads to a nuclear magnetization. The macroscopic magnetization M of the nuclei in the sample is the sum of individual nuclear magnetic moments $\boldsymbol{\mu}_i$ per volume V

$$\boldsymbol{M} = \sum_i^n \frac{\boldsymbol{\mu}_i}{V} \tag{6.6}$$

The expectation value of the z component of M in the external field $\boldsymbol{B}_0$ (z axis chosen in the direction of $\boldsymbol{B}_0$) is

$$M_z = \sum_i^n \frac{\mu_{iz}}{V} = N\gamma \langle I_z \rangle \tag{6.7}$$

where N is the density of nuclear spins, and $\langle I_z \rangle$ is the average nuclear spin polarization calculated above. Substitution of $\langle I_z \rangle$ from Eq. (6.4) gives for the magnetization in thermal equilibrium

$$M_0 = N\gamma \frac{\gamma \hbar^2 I(I+1)}{3k_B T} B_0 \tag{6.8}$$

The factor multiplying B_0 is the nuclear spin susceptibility divided by the vacuum permittivity μ_0

$$\chi_K = \frac{N\gamma^2 \hbar^2 I(I+1)}{3k_B T} \mu_0 \tag{6.9}$$

Since the nuclear gyromagnetic ratio is approximately one thousand times smaller in magnitude than the electron gyromagnetic ratio, the nuclear spin susceptibility is approximately six orders of magnitude smaller than electronic susceptibilities (because they are proportional to γ squared). The nuclear spin susceptibility follows a $1/T$ temperature dependence (Curie law).

Resonance absorption of radiofrequency waves

Irradiation with electromagnetic waves having frequency $\omega = \omega_L$ causes resonant absorption. Because of the unequal occupation of the lower and higher energy levels, somewhat more quanta are absorbed than emitted. There is competition between the high-frequency radiation, which tries to establish an equal distribution, and the lattice, which tries to establish a Boltzmann distribution.

NMR is suitable for investigation of solids for the following reasons: First, the resonant frequency is not exactly the Larmor frequency caused by the external field but is slightly shifted. The reason for the shift lies in the (paramagnetic) enhancement or (diamagnetic) reduction of the external field by the electrons surrounding the nucleus. Precise measurement of these shifts provides information about the surroundings of the nucleus. Secondly, the resonance line exhibits a width which is associated with relaxation processes in the solid. One gains information about the microscopic properties through measurements of these NMR linewidths or relaxation times.

The set-up of an NMR experiment is shown schematically in Fig. 6.1. The

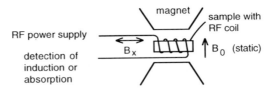

Fig. 6.1 Schematic diagram of an NMR experiment.

sample is placed in a static external B field B_0. In addition, a time-dependent radiofrequency field B_{RF} is applied perpendicular to B_0. The radiofrequency field has the form

$$B_{RF} = 2B_1 \cos(\omega t) \tag{6.10}$$

Such a field can be decomposed into two oppositely rotating magnetic fields with frequencies $+\omega$ and $-\omega$. When one of these is near the resonant frequency (e.g. $+\omega$) the contribution of the other (e.g. $-\omega$) is negligible. Thus the total field is composed of a static field B_0 in the z direction and a rotating field B_1 in the xy plane, i.e.

$$B_x = B_1 \cos(\omega t), \quad B_y = B_1 \sin(\omega t), \quad B_z = B_0 \tag{6.11}$$

If one varies either the frequency or the static field, one eventually reaches the resonant condition which shows up as an absorption in the primary coil (Purcell, *et al.*, 1946) or induction in a secondary coil (Bloch, *et al.*, 1946).

6.2 Classical Treatment of NMR (Bloch Equations)

We have seen that applying an external B_0 field to a sample of nuclear spins establishes a magnetization M. The behavior of this magnetization under the simultaneous influence of the B_0 field, the radiofrequency field, and the surroundings will be discussed in the following.

Free motion equation

In an external B field, a magnetic moment MV is subject to a torque of magnitude $(MV \times B)$ which leads to a change of the total angular momentum I

$$\frac{dI}{dt} = (MV \times B) \tag{6.12}$$

or, because of $M = \gamma \sum_i^n I_i/V = \gamma I/V$

$$\frac{dM}{dt} = \gamma(M \times B) \tag{6.13}$$

The solution of Eq. (6.13) gives a precession of the magnetization M around the magnetic field B; the precession frequency is $\omega_L = -\gamma B$.

Relaxation

In thermal equilibrium the magnetization in an external field is

$$M_z = M_0 \qquad M_x = M_y = 0 \tag{6.14}$$

Deviations from these values relax with certain time constants (T_1, T_2) to the thermal equilibrium values. One makes the following *ansatz* (Bloch, 1946)

$$\frac{dM_z}{dt} = \frac{M_0 - M_z}{T_1}$$
$$\frac{dM_x}{dt} = -\frac{M_x}{T_2} \tag{6.15}$$
$$\frac{dM_y}{dt} = -\frac{dM_y}{T_2}$$

T_1 is called the *longitudinal* or *spin–lattice relaxation time*. It is a characteristic time during which the energy from the nuclear spin system (repopulation of M levels) is transferred to the lattice. A graphical interpretation of T_1 is shown in Fig. 6.2. If an unmagnetized sample ($M = 0$) is brought into a field B_0 at $t = 0$, then the magnetization rises with the time constant T_1 from the value 0 to the equilibrium value M_0

$$M_z(t) = M_0[1 - \exp(-t/T_1)] \tag{6.16}$$

T_2 is called the *transverse* or *spin–spin relaxation time*. It has a totally different meaning than T_1. T_2 is a measure of the time during which individual moments which contribute to M_x and M_y stay in phase. Expressed differently, if at a certain time, a number of nuclear moments point in the same direction (e.g. x direction), these will dephase with time because of slight differences in the precession frequencies. T_2 is the characteristic time for this dephasing process. For this reason it is also called phase relaxation.

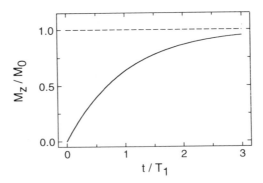

Fig. 6.2 Time dependence of the magnetization M_z following instantaneous turning-on of a magnetic field.

The designation spin–spin relaxation time for T_2 arises from the fact that differences in precession frequency are often caused by the spin–spin-interaction of the observed spin with its neighbors. The designation transverse (or longitudinal for T_1) relaxation time arises directly from the definition (Eq. (6.15)) since T_2 is related to the transverse (perpendicular to B_0) and T_1 to the longitudinal (parallel to B_0) component of M.

Of course there are no such processes for the longitudinal component which change the phase relationship, since M_z does not precess. On the other hand spin–lattice processes can also contribute to T_2.

Combination of the free motion equations (Eq. (6.13)) with the relaxation terms (Eq. (6.15)) leads to the Bloch equations in the laboratory system

$$\frac{dM_z}{dt} = \gamma(\boldsymbol{M} \times \boldsymbol{B})_z + \frac{M_0 - M_z}{T_1}$$

$$\frac{dM_x}{dt} = \gamma(\boldsymbol{M} \times \boldsymbol{B})_x - \frac{M_x}{T_2} \qquad (6.17)$$

$$\frac{dM_y}{dt} = \gamma(\boldsymbol{M} \times \boldsymbol{B})_y - \frac{M_y}{T_2}$$

For the solution of the Bloch equations one usually goes into the rotating coordinate system in which one obtains particularly simple and intuitive solutions.

Transformation in a rotating coordinate system

We consider a coordinate system that rotates around the B_0 axis with the same frequency as the radiofrequency field. For the observer in the rotating system (x', y', z') the B_1 field is fixed. In general, we take x' in the direction of B_1.

Transformation of the equations (6.17) into the rotating system causes the differentials to include an additional term (Coriolis effect). In general, a given vector F is transformed in the following way

$$\left(\frac{dF}{dt}\right)_{rot} = \left(\frac{dF}{dt}\right)_{fixed} - (\omega \times F) \tag{6.18}$$

where 'rot' and 'fixed' designate rotating and fixed coordinate systems respectively.

Thus the equations of motion (without relaxation) in the rotating system are

$$\left(\frac{dM}{dt}\right)_{rot} = \gamma(M \times B) - (\omega \times M)$$

$$= \gamma\left[M \times \left(B + \frac{\omega}{\gamma}\right)\right] = \gamma(M \times B_{eff}) \tag{6.19}$$

where B is the external B_0 field plus the radiofrequency field B_1, and ω the rotation frequency. B_{eff} is defined by

$$B_{eff} = B + \frac{\omega}{\gamma} \tag{6.20}$$

With unit vectors $e_{x'}$ and e_z in the rotating system, Eq. (6.20) can be written

$$B_{eff} = (B_0 - B_\omega)e_z + B_1 e_{x'} \quad \text{where } B_\omega = -\frac{\omega}{\gamma} \tag{6.21}$$

The different contributions to B_{eff} are shown in Fig. 6.3.

By transformation into the rotating system, the time dependence of the B_1 field was eliminated; the resulting equations of motion (6.19) have the same form as (6.13). Thus the solution is a precession of the magnetization around B_{eff}; thereby the z component of the magnetization is changed. M_z is changed most effectively if $\omega = \omega_L = -\gamma B_0$, i.e. if the frequency of the B_1 field is equal to the Larmor frequency. Then B_{eff} is identical to B_1 and the magnetization precesses around B_1.

One sees from Fig. 6.3 that the magnetization that originally pointed in the z direction can be rotated into the y' direction or the $-z$ direction by applying B_1 for appropriate time intervals. One speaks of a 90° or 180° pulse. In this way, a well-defined non-equilibrium state can be created. One

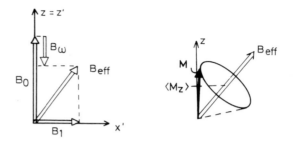

Fig. 6.3 Composition of B_{eff} in the rotating coordinate system. The precession of M around the effective magnetic field in the rotating system is shown on the right side.

can measure the re-establishment of the equilibrium state and consequently T_1 and T_2. This technique is used in many NMR investigations.

Adding into Eq. (6.19) the relaxation terms (Eq. (6.15)) and substituting

$$\omega_{L,1} = -\gamma B_1 \quad \text{and} \quad \omega_{L,0} = -\gamma B_0 \tag{6.22}$$

leads to the Bloch equations in the rotating system

$$\frac{d\widetilde{M}_x}{dt} = (\omega - \omega_{L,0})\widetilde{M}_y - \frac{\widetilde{M}_x}{T_2}$$
$$\frac{d\widetilde{M}_y}{dt} = -(\omega - \omega_{L,0})\widetilde{M}_x - \omega_{L,1}\widetilde{M}_z - \frac{\widetilde{M}_y}{T_2} \tag{6.23}$$
$$\frac{d\widetilde{M}_z}{dt} = \omega_{L,1}\widetilde{M}_y - \frac{\widetilde{M}_z - M_0}{T_1}$$

where the superscript $\sim$ indicates the rotating system.

In the following we give the solutions to Eq. (6.23) for the case of slow passage. If B_0 or ω are changed so slowly that the system remains in equilibrium at all times, one obtains

$$\frac{d\widetilde{M}_x}{dt} = \frac{d\widetilde{M}_y}{dt} = \frac{d\widetilde{M}_z}{dt} = 0 \tag{6.24}$$

In this case, Eq. (6.23) reduces to three algebraic equations with three unknowns. The solutions are

$$\widetilde{M}_x = \frac{(\omega - \omega_{L,0})\gamma B_1 T_2^2}{1 + \gamma^2 B_1^2 T_1 T_2 + (\omega - \omega_{L,0})^2 T_2^2} M_0$$

$$\widetilde{M}_y = \frac{\gamma B_1 T_2}{1 + \gamma^2 B_1^2 T_1 T_2 + (\omega - \omega_{L,0})^2 T_2^2} M_0 \qquad (6.25)$$

$$\widetilde{M}_z = \frac{1 + (\omega - \omega_{L,0})^2 T_2^2}{1 + \gamma^2 B_1^2 T_1 T_2 + (\omega - \omega_{L,0})^2 T_2^2} M_0$$

For $B_1 = 0$, i.e. for vanishing RF fields, the solutions, Eq. (6.25), become the equilibrium values in the static B_0 field, $\widetilde{M}_z = M_0$ and $\widetilde{M}_x = \widetilde{M}_y = 0$. The same is true for the case where the terms with $(\omega - \omega_{L,0})$ become large compared to other terms, i.e. far away from resonance. However, close to the resonance, as can be seen in Eq. (6.25), the z component is reduced and partially rotated into the $x'y'$ plane (see Fig. 6.4). This part of the magnetization is not constant and can be detected as an alternating voltage induced in a fixed coil (see Section 6.3). Using a phase-sensitive amplifier, the $\widetilde{M}_x$ and $\widetilde{M}_y$ components can be detected separately.

As can be seen from Eq. (6.25), $\widetilde{M}_x$ displays a dispersion curve and $\widetilde{M}_y$ an absorption curve (Fig. 6.5). This can be seen better if one writes

$$\widetilde{M}_x = \gamma B_1 M_0 \frac{\omega - \omega_{L,0}}{(\omega - \omega_{L,0})^2 + \Gamma^2/4} \qquad (6.26)$$

$$\widetilde{M}_y = \frac{\gamma B_1 M_0}{T_2} \frac{1}{(\omega - \omega_{L,0})^2 + \Gamma^2/4}$$

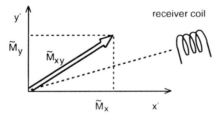

Fig. 6.4 Magnetization in the $x'y'$ plane in the rotating coordinate system. The magnetization rotates at frequency ω with respect to the fixed coil.

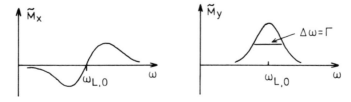

Fig. 6.5 Dispersion signal of $\widetilde{M}_x$ and absorption signal of $\widetilde{M}_y$.

with

$$\Gamma = \frac{2}{T_2}\sqrt{1 + \gamma^2 B_1^2 T_1 T_2} \qquad (6.27)$$

The linewidth Γ is given in the two limiting cases as follows:
(a) For weak radiofrequency fields, i.e. $\gamma^2 B_1^2 T_1 T_2 \ll 1$,

$$\Delta\omega = \Gamma = \frac{2}{T_2} \qquad (6.28)$$

In this case the linewidth is dependent only on T_2. From the measurement of the linewidth, the spin–spin relaxation time can be determined.
(b) For strong radiofrequency fields, i.e. $\gamma^2 B_1^2 T_1 T_2 \gg 1$,

$$\Delta\omega = \Gamma = 2\gamma B_1 \sqrt{\frac{T_1}{T_2}} \qquad (6.29)$$

For a strong radiofrequency field, Γ depends also on the strength of the B_1 field and therefore can be varied experimentally (power broadening).

6.3 Experimental Methods

Since all stable nuclei with $I \geq 1/2$ are suitable, NMR is a versatile and common method. A selection of NMR nuclei with their important parameters is given in Table 6.1. Aside from stable nuclei, radioactive nuclei can be used in experiments in special cases. We return to this point in section 6.7.

Two set-ups are most commonly used for NMR experiments:
(a) In the stationary or continuous-wave (CW) method, a continuous but weak radiofrequency field ($B_1 \approx 10^{-7}$ T) acts on the sample.
(b) In pulsed NMR, the magnetization is turned out of the z direction by a short but strong RF pulse ($B_1 \approx 10^{-3}$ T), and after the RF pulse ends, the free precession and relaxation is observed in a receiver coil.

Continuous-wave methods are increasingly being replaced by pulsed methods, since pulsed NMR offers larger sensitivity and more flexibility.

Table 6.1 Properties of some selected NMR nuclei (values adopted from Lederer and Shirley, 1978).

Isotope	Natural abundance (%)	Spin I	μ (μ_N)	Q (barn)	NMR frequency (MHz T^{-1})
^{1}H	99.985	1/2	+2.793	0	42.576
^{2}H	0.0148	1	+0.857	+0.00288	6.532
^{7}Li	92.5	3/2	+3.256	−0.4	16.545
^{13}C	1.11	1/2	+0.702	0	10.701
^{19}F	100	1/2	+2.629	0	40.076
^{27}Al	100	5/2	+3.642	+0.15	11.104
^{31}P	100	1/2	+1.132	0	17.256
^{35}Cl	75.77	3/2	+0.822	−0.082	4.177
^{63}Cu	69.2	3/2	+2.223	−0.21	11.296
^{105}Pd	22.2	5/2	−0.642	+0.8	1.957
^{127}I	100	5/2	+2.813	−0.79	8.576
^{195}Pt	33.8	1/2	+0.609	0	9.283
^{207}Pb	22.1	1/2	+0.593	0	9.040

6.3.1 Continuous-Wave Method

Here we will describe only the set-up with a single coil (Purcell method). In principle of course a separation of the transmitter and receiver coil is possible (Bloch method). The basic idea in both methods is the same: the magnetization of the nuclear spin system precessing in the xy plane induces a voltage in a coil. The voltage can be calculated in the following way. The B field induced in the sample has the value

$$B = \mu_0 M \tag{6.30}$$

The associated magnetic flux Φ in the coil is

$$\Phi = A \eta B_x \tag{6.31}$$

where A is the area of the coil times the number of turns, and η is the filling factor, which is the ratio of the effective probe volume to the total volume of the coil.

The back-transformation of the magnetization from the rotating into the fixed coordinate system yields for the x component

$$M_x(t) = \widetilde{M}_x \cos \omega t - \widetilde{M}_y \sin \omega t \tag{6.32}$$

Combining Eq. (6.30)–(6.32) yields for the induced voltage

$$U_r = -\frac{d\Phi}{dt} = A\eta\mu_0\omega(\widetilde{M}_x \sin \omega t + \widetilde{M}_y \cos \omega t) \qquad (6.33)$$

By a compensation circuit (Purcell bridge, see Fig. 6.6) it is possible to compensate the transmitter voltage so that following the bridge a voltage proportional to U_r is present. One can see from Eq. (6.33) that U_r represents a superposition of the dispersive ($\widetilde{M}_x$) and absorptive ($\widetilde{M}_y$) parts. By using a phase-sensitive detector (Section 6.3.2) it is possible to separate the two parts. By choosing the lock-in phase $\delta = 0°$ or $\delta = 90°$, one obtains the absorption or dispersion signal respectively. A block diagram of the NMR electronics is shown in Fig. 6.7.

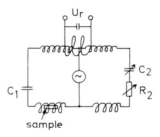

Fig. 6.6 Purcell bridge for the compensation of the transmitter voltage. The bridge is balanced off-resonance.

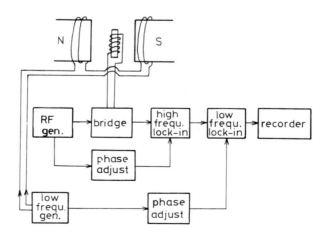

Fig. 6.7 Block diagram of the NMR apparatus for the continuous-wave (CW) method.

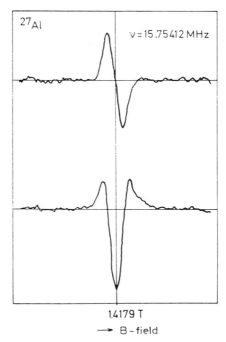

Fig. 6.8 Absorption and dispersion signal of the ^{27}Al resonance. The differential signals are shown here (Stachel, 1982).

The RF generator is connected to the transmitter coil through the Purcell bridge. The output of the bridge is processed further in the RF lock-in amplifier which receives a reference signal from the RF generator. Depending on the choice of phase shift, one obtains an output DC voltage proportional to either $\widetilde{M}_y$ ($\delta = 0°$) or $\widetilde{M}_x$ ($\delta = 90°$).

In order to improve the signal-to-noise ratio, one uses the lock-in principle a second time. One modulates the B_0 field with a low-frequency oscillation having amplitude small compared to the width of the NMR line. Only signals that have the phase and frequency of the magnetic field modulation pass through the second lock-in amplifier.

The resonance line is detected by a slow variation of the RF frequency or the B_0 field. An example of the absorption and dispersion signal is shown in Fig. 6.8.

6.3.2 Lock-in Amplifier

Detection of a small signal in the presence of large noise is a common problem in experimental science. The lock-in principle is a method to improve the signal-to-noise ratio. The voltage U_S which is to be measured is periodically modulated by some physical parameter p on which U_S depends. An example from the preceding subsection is the induced voltage appearing after the Purcell bridge as a function of the magnetic field. For that the parameter p is periodically varied with the reference frequency ω_{ref} around a set value p_0

$$p = p_0 + \Delta p \sin \omega_{\text{ref}} t \qquad (6.34)$$

With a Taylor expansion, one obtains the behavior of the measured signal U_S

$$U_S(p, t) = U_S(p_0) + \left.\frac{\partial U_S}{\partial p}\right|_{p_0} \Delta p \sin \omega_{\text{ref}} t + \cdots \qquad (6.35)$$

One sees that the time variation of the measured value U_S is locked to the reference frequency.

If one now sends the noisy measured signal through a band-pass filter centered at the reference frequency, one obtains the time-varying part of the output signal and only the noise that passes the filter. The narrower the filter bandwidth, the higher the signal-to-noise ratio.

A block diagram is shown in the upper part of Fig. 6.9. In the lower part of that figure, the voltage is shown for different phases between the signal U_S and the reference U_{ref}.

One sees that the first state works as a multiplier, i.e. it produces the product $U_S \otimes U_{\text{ref}}$ at the output. The product voltage $U_\otimes$ can be easily calculated if the Fourier transform of the square wave U_{ref} is used, and the phase shift between U_{ref} and the switches S_1, S_2 is called δ.

In addition, a time-varying noise signal $U_R(t)$ is added to the measured signal (Eq. (6.35))

$$U_\otimes = [U_S(p, t) + U_R(t)]U_{\text{ref}}$$

$$= \left[U_S(p_0) + U_R(t) + \left.\frac{\partial U_S}{\partial p}\right|_{p_0} \Delta p \sin(\omega_{\text{ref}} t) + \cdots \right]$$

$$\times \left[\sin(\omega_{\text{ref}} t + \delta) + \frac{1}{3}\sin[3(\omega_{\text{ref}} t + \delta)] + \cdots \right] \qquad (6.36)$$

Limiting ourselves to the first term in the Fourier series of U_{ref}, we obtain

$$U_\otimes \approx U_S(p_0) \sin(\omega_{\text{ref}} t + \delta) + U_R(t) \sin(\omega_{\text{ref}} t + \delta)$$

$$- \left.\frac{\partial U_S}{\partial p}\right|_{p_0} \frac{\Delta p}{2} \cos(2\omega_{\text{ref}} t + \delta) + \left.\frac{\partial U_S}{\partial p}\right|_{p_0} \frac{\Delta p}{2} \cos \delta \qquad (6.37)$$

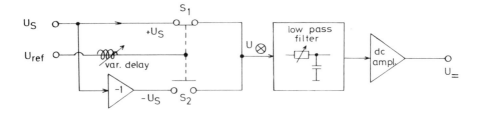

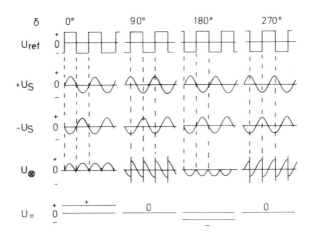

Fig. 6.9 Top: Block diagram of a lock-in amplifier. If the reference voltage U_{ref} is positive the switch S_1 is closed. In the opposite case, S_2 is closed. Bottom: Voltage output for different phase relationships between signal voltage U_S and reference voltage U_{ref}.

This voltage goes to a low-pass filter which ideally passes only the DC voltage portion of the signal. Such a DC voltage can be seen in the last term of Eq. (6.37); this voltage is a measure of the first derivative of our signal relative to the modulated parameter p and can be maximized or made zero with appropriate choice of the phase shift δ (see $U_=$ in Fig. 6.9). However, the second term of Eq. (6.37) contains time-independent parts originating from noise seen at the reference frequency. The contribution of noise at ω_{ref} is thus passed. However, it is usually small compared to the desired signal. Consequently the whole circuit has the effect of a phase-sensitive rectifier with a filtering at the reference frequency whose Q factor is determined by the low-pass filter. By this method one can reach extremely high Q factors with typical values $Q_{filt} = 10^8$.

6.3.3 Pulsed NMR

In Fig. 6.3 we showed that the magnetization can be turned out of the z direction by a radiofrequency B_1 field. The special case for which the RF field is exactly the Larmor frequency is shown in Fig. 6.10. In this case $B_{\text{eff}} = B_1$ (see Eq. (6.21)) and the magnetization precesses around the x' axis. The precession angle is

$$\alpha = \gamma B_1 t_p \tag{6.38}$$

where t_p is the duration of the RF pulse. The precession angle therefore can be chosen by the duration of the RF pulse.

After the 90° pulse the magnetization, originally lying in the z direction, is now in the y' direction. If the RF field is now switched off the magnetization will perform a free precession in the xy plane and finally return to the equilibrium position via relaxation processes (T_1 and T_2 processes). This is called free-induction decay.

The observation of the precession is done in a spatially fixed receiver coil which usually is the same coil as that used for transmission. The switching of the coil from transmission to receiving, and other experimental processes, are under computer control. The voltage induced in the coil is again determined by Eqs. (6.30)–(6.33). However, there is an important difference compared to the CW method: in CW-NMR the different resonance lines are observed one after the other by variation of the RF frequency (or B_0 field); in pulsed NMR the different precession frequencies are observed simultaneously. One can say that the CW method works in frequency space and pulsed NMR in time space.

If the static B_0 field is not the same for all nuclei (e.g. external B_0 field

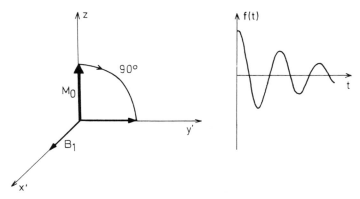

Fig. 6.10 Representation of a 90° pulse (on the left) and the subsequent free-induction decay (on the right).

plus chemical shifts) the effective B field B_{eff} in the rotating frame is also different. Only a fraction of the nuclei have $B_{eff} = B_1$ and only these nuclei perform an exact 90° rotation during the time t_p. The other nuclear spins do not end up exactly in the xy plane and therefore contribute only partially to the free-induction decay. In order that nuclei contribute with full amplitude to the free-induction decay, it is necessary that $|\omega_L - \omega| \ll \gamma B_1$. In other words, the angle between B_{eff} and B_1 must be very small (compare Fig. 6.3). Then one obtains, with $\alpha = \pi/2 = 90°$ in Eq. (6.38),

$$2\pi\Delta := |\omega_L - \omega| \ll \gamma B_1 = 2\pi/(4t_p) \tag{6.39}$$

i.e.

$$t_p \ll 1/(4\Delta)$$

where Δ is the frequency region around the carrier frequency ω within which the free-induction decay can be observed with full amplitude (Fig. 6.11). The underlying frequencies and their linewidths can be observed by Fourier transformation of the time spectrum.

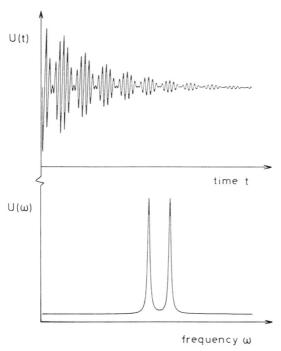

Fig. 6.11 Interference pattern of two free-induction decays. In the lower part of the figure the Fourier spectrum is shown (Holz and Knüttel, 1983).

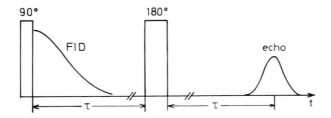

Fig. 6.12 Principles of the spin-echo method. After the 90° pulse the transverse magnetization decays by free induction (FID). After time τ all spins are reversed by a 180° pulse. After another time τ the echo occurs, i.e. all spins are again in phase.

6.3.4 Spin–Echo Method

The decay of the transverse magnetization after the 90° pulse is often caused by variations in the local magnetic field for the different nuclei. The origin of these variations may be inhomogeneity of the externally applied magnetic field or internal inhomogeneities, e.g. dipolar fields of neighboring nuclei. Time-independent inhomogeneities can be eliminated by the spin–echo method.

The basic idea of the spin–echo method is displayed in Fig. 6.12. After a 90° pulse, the nuclear spins in a slightly inhomogeneous magnetic field precess with slightly different frequencies and therefore dephase with time. The macroscopic magnetization of the nuclear spin system therefore is lost. However, application of a 180° pulse at some time τ causes all spin orientations to be reversed and the phases of the spins to come back together at the same rate that they dephased prior to the 180° pulse. After time 2τ (assuming the duration of the 180° pulse is negligible) all spins are again in phase, i.e. the macroscopic magnetization is restored, and one obtains an echo.

However, if the nuclei have changed positions during the time 2τ and have moved to a region with a different magnetic field, or if the field has varied in time, then not all spins come together after 2τ, and the strength of the echo is diminished. Therefore it is possible to separate the interesting effects from the trivial effects of field inhomogeneities by the spin–echo method.

6.4 Chemical Shift

The most important application of NMR is probably not in solid state physics but in chemistry and biochemistry where the chemical shift is observed. The chemical shift arises because the electronic environment of the nucleus shifts the resonance in a characteristic way.

The chemical shift is the small difference between the B field at the nuclear site and the external field B_0

$$B_{nuclear} = (1 - \sigma) B_0 \qquad (6.40)$$

The chemical shift σ is divided into two parts, a diamagnetic contribution σ_D that decreases the field, and a paramagnetic contribution σ_P that increases the field. With this separation, one obtains

$$B_{nuclear} = (1 - \sigma_D + \sigma_P) B_0 \qquad (6.41)$$

The shifts are of order $\Delta\omega/\omega \approx 10^{-5}$ and can be measured easily with high-resolution NMR (typical resolution $\Delta\omega/\omega \approx 10^{-7}$).

As an example, in Fig. 6.13 the proton resonance of ethanol is shown; one recognizes three resonances which originate from hydrogen with different chemical environments. The chemical shift is characteristic of the molecular bonding of the atoms (see Fig. 6.14) and therefore can be used to identify molecular structures.

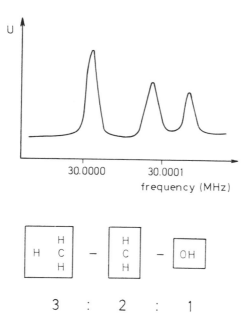

Fig. 6.13 Proton NMR signal of ethanol. The ethanol structure is shown in the lower part of the figure. One recognizes three non-equivalent hydrogen positions.

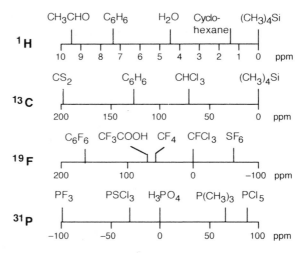

Fig. 6.14 Typical chemical shifts (in ppm) for the NMR nuclei ^{1}H, ^{13}C, ^{19}F, and ^{31}P (Shaw, 1976).

It is important to note that the chemical shift is well defined only for diamagnetic molecules, i.e. molecules which have no unpaired electrons and for which $\langle L_z \rangle = 0$.

The chemical shift contributions σ_D and σ_P have the following qualitative origins (a more detailed description will be given below):

(a) Diamagnetic shift σ_D. Turning on the magnetic field induces a shielding current in the electronic shell of the molecule. This current reduces the external field (Lenz rule).
(b) Paramagnetic shift σ_P. Application of an external field induces magnetic dipole moments that align with the magnetic field.

We now undertake a more quantitative discussion for a NMR nucleus in a diamagnetic environment where it senses only the external field $\boldsymbol{B}_0$ in zeroth order. In higher order, an indirect influence of the electronic shell shows up in the sense that the magnetic field changes the electronic shell, and then the changed electronic shell produces an additional magnetic field at the nuclear site. Since this is only a weak effect it can be treated by perturbation theory.

In addition to the external field $\boldsymbol{B}_0$, an electron near the nucleus experiences a field $\boldsymbol{B}_\mu$ which originates from the nuclear dipole moment $\boldsymbol{\mu}$. The two fields can be represented by vector potentials

$$\boldsymbol{A}_0 = \frac{1}{2}(\boldsymbol{B}_0 \times \boldsymbol{r})$$

$$A_\mu = \frac{\mu_0}{4\pi r^3}(\boldsymbol{\mu} \times \boldsymbol{r}) \tag{6.42}$$

$$A = A_0 + A_\mu = \left(\frac{1}{2}B_0 + \frac{\mu_0}{4\pi r^3}\boldsymbol{\mu}\right) \times \boldsymbol{r}$$

The total Hamiltonian operator has the form

$$\begin{aligned}\mathcal{H} &= \frac{1}{2m_e}(\boldsymbol{p} + e\boldsymbol{A})^2 - eU \\ &= -\frac{\hbar^2}{2m_e}\nabla^2 + \frac{e}{2m_e}(\boldsymbol{p}\cdot\boldsymbol{A} + \boldsymbol{A}\cdot\boldsymbol{p}) + \frac{e^2}{2m_e}\boldsymbol{A}^2 - eU \end{aligned} \tag{6.43}$$

where U is the molecular potential. The term $\boldsymbol{p}\cdot\boldsymbol{A}$ must be considered more closely since the differential operator $\boldsymbol{p} = -i\hbar\nabla$ acts on both $\boldsymbol{A}$ and the wave function. Applying the product rules yields

$$\boldsymbol{p}\cdot\boldsymbol{A} = -i\hbar(\nabla\cdot\boldsymbol{A}) + \boldsymbol{A}\cdot\boldsymbol{p} \tag{6.44}$$

Since we have chosen in Eq. (6.42) the Coulomb gage, i.e. $\nabla\cdot\boldsymbol{A} = 0$, we obtain

$$\boldsymbol{p}\cdot\boldsymbol{A} = \boldsymbol{A}\cdot\boldsymbol{p} \tag{6.45}$$

The part $-(\hbar^2/2m_e)\nabla^2 - eU$ in Eq. (6.43) is the unperturbed Hamiltonian operator. The remainder, which describes the perturbation, is

$$\mathcal{H}' = \frac{e}{2m_e}[2(\boldsymbol{A}\cdot\boldsymbol{p}) + e\boldsymbol{A}^2] \tag{6.46}$$

$2(\boldsymbol{A}\cdot\boldsymbol{p})$ can be written in the following way

$$\begin{aligned} 2(\boldsymbol{A}\cdot\boldsymbol{p}) &= \left[\left(B_0 + \frac{\mu_0}{2\pi r^3}\boldsymbol{\mu}\right) \times \boldsymbol{r}\right]\cdot\boldsymbol{p} \\ &= \left(B_0 + \frac{\mu_0}{2\pi r^3}\boldsymbol{\mu}\right)\cdot(\boldsymbol{r} \times \boldsymbol{p}) \\ &= \left(B_0 + \frac{\mu_0}{2\pi r^3}\boldsymbol{\mu}\right)\cdot\boldsymbol{l} \end{aligned} \tag{6.47}$$

Nuclear Magnetic Resonance (NMR)

If summed over all electrons j of the shell and if one uses the Bohr magneton $\mu_B = e\hbar/2m_e$ the perturbation operator becomes

$$\mathcal{H}' = \sum_j \frac{\mu_B}{\hbar}\left(B_0 + \frac{\mu_0}{2\pi r_j^3}\mu\right) \cdot l_j + \frac{m_e \mu_B^2}{2\hbar^2}\left[\left(B_0 + \frac{\mu_0}{2\pi r_j^3}\mu\right) \times r_j\right]^2 \quad (6.48)$$

Since we are interested only in the interaction of the nucleus with changes induced in the electronic shell by the external field B_0, only terms containing both B_0 and μ have to be considered. Such a term is obtained by multiplying out the last terms in Eq. (6.48). Since this expression will give the diamagnetic shielding, we add the index D

$$\mathcal{H}'_D = \sum_j \frac{m_e \mu_B^2}{2\hbar^2} 2(B_0 \times r_j) \frac{\mu_0}{2\pi r_j^3}(\mu \times r_j) \quad (6.49)$$

Using first-order perturbation theory and the fact that μ has a nonzero expectation value only in the direction of B_0 (z direction) leads to

$$E'_D = \mu_z B_0 \left[\frac{m_e \mu_0 \mu_B^2}{2\pi \hbar^2} \sum_j \int \psi_j^* \frac{x_j^2 + y_j^2}{r_j^3} \psi_j \, d^3r\right]$$

$$= -\mu_z B_0(-\sigma_D) \quad (6.50)$$

with

$$\sigma_D = \frac{m_e \mu_0 \mu_B^2}{2\pi \hbar^2} \sum_j \int \psi_j^* \frac{x_j^2 + y_j^2}{r_j^3} \psi_j \, d^3r \quad (6.51)$$

One sees that σ_D is always positive, and therefore, according to Eq. (6.40), always decreases the magnetic field. For hydrogen, $\sigma_D \approx 3 \times 10^{-5}$, and for lead, $\sigma_D \approx 1 \times 10^{-2}$. For s electrons, σ_D is given by the expression derived by Lamb (1941)

$$\sigma_D(s) = \frac{m_e \mu_0 \mu_B^2}{2\pi \hbar^2} \sum_j \int \psi_j^* \frac{2}{3r_j} \psi_j \, d^3r$$

$$= \frac{\mu_0}{4\pi} \frac{4 m_e \mu_B^2}{3\hbar^2} \left\langle \frac{1}{r} \right\rangle \quad (6.52)$$

In this approximation, σ_D depends only on the radial distribution of electrons in the ground state.

In first-order perturbation theory there are no other bilinear terms in μ

and B_0. However, one obtains an additional bilinear term in second order. The general expression of the perturbation energy in second order is

$$\Delta = \sum_n \frac{\langle 0|\mathcal{H}'|n\rangle\langle n|\mathcal{H}'|0\rangle}{E_n - E_0} \quad (6.53)$$

where summation over all excited states $|n\rangle$ is performed. Limiting consideration to terms bilinear in μ and B_0 leads to

$$E'_P = \sum_{n,j,k} \frac{\langle 0|\frac{\mu_B}{\hbar}B_0 \cdot l_k|n\rangle\langle n|\frac{\mu_B}{\hbar}\frac{\mu_0}{2\pi r_j^3}\mu \cdot l_j|0\rangle}{E_n - E_0} \quad (6.54)$$

Since only the z component of μ is observable, one obtains

$$E'_P = -\mu_z B_0 \left[\frac{\mu_0 \mu_B^2}{2\pi\hbar^2} \sum_n \frac{1}{E_n - E_0} \langle 0|\sum_k l_{kz}|n\rangle\langle n|\sum_j \frac{l_{jz}}{r_j^3}|0\rangle \right]$$

$$= -\mu_z B_0 \sigma_P \quad (6.55)$$

with the paramagnetic enhancement

$$\sigma_P = \frac{\mu_0 \mu_B^2}{2\pi\hbar^2} \sum_n \frac{1}{E_n - E_0} \langle 0|\sum_k l_{kz}|n\rangle\langle n|\sum_j \frac{l_{jz}}{r_j^3}|0\rangle \quad (6.56)$$

The calculation of σ_P requires knowledge of the excited states of the atoms or molecules in which the nucleus is embedded; therefore it is a complicated computation. Intuitively, σ_P arises because the B field mixes an excited state having $\langle L_z \rangle \neq 0$ with the molecular ground state having $\langle L_z \rangle = 0$. Therefore the system attains a paramagnetic moment.

6.5 Knight Shift in Metals

The Knight shift is the shift of the nuclear resonance by *polarized conduction electrons*. It is named for W. D. Knight who discovered this resonance shift in copper in 1949 (Knight, 1949). The polarization of the conduction electrons results from the raising and lowering of the spin-up and spin-down bands in an external field, respectively. It has the same origin as the Pauli susceptibility. We first discuss this phenomenon and then describe the effect of the polarized electrons on the nuclear spin.

Pauli susceptibility

In the framework of the free electron gas, the density of states $D(E)$ is

$$D(E) = \frac{3N}{2E_F^{3/2}}\sqrt{E} \tag{6.57}$$

where N is the electron density and E_F the Fermi energy. Without an external field the states are occupied equally with spin-up and spin-down electrons. When the field B_0 is turned on, the bands are shifted (see Fig. 6.15). The band having magnetic moment parallel to B_0 is lowered by $\mu_B B_0$ and the band with magnetic moment antiparallel to B_0 is raised by the same energy. Since both bands are filled to the Fermi energy, there is an excess of electrons having μ_S parallel to B_0, and consequently the sample has a nonzero magnetic moment and electronic spin polarization.

The density of electrons with μ_S parallel to B_0 is

$$\begin{aligned}N_+ &= \frac{1}{2}\int_{-\mu_B B_0}^{\infty} f(E) D(E + \mu_B B_0)\, dE \\ &\approx \frac{1}{2}\int_0^{\infty} f(E) D(E)\, dE + \frac{1}{2}\mu_B B_0 D(E_F)\end{aligned} \tag{6.58}$$

where $f(E)$ is the Fermi distribution function. Eq. (6.58) is valid for $k_B T \ll E_F$. In addition it is assumed that the shift $\mu_B B_0$ is small so that $D(E)$ can be approximated by $D(E_F)$ in the region between E_F and $E_F + \mu_B B_0$. Analogously the density of electrons with μ_S antiparallel to B_0 is

$$\begin{aligned}N_- &= \frac{1}{2}\int_{\mu_B B_0}^{\infty} f(E) D(E - \mu_B B_0)\, dE \\ &\approx \frac{1}{2}\int_0^{\infty} f(E) D(E)\, dE - \frac{1}{2}\mu_B B_0 D(E_F)\end{aligned} \tag{6.59}$$

From these relations, one obtains for the magnetization

$$M = \mu_B(N_+ - N_-) = \mu_B^2 B_0 D(E_F) = \frac{3N\mu_B^2}{2k_B T_F} B_0 \tag{6.60}$$

where $T_F = E_F/k_B$ is the Fermi temperature. This yields for the Pauli susceptibility

$$\chi_{\text{Pauli}} = \frac{\mu_0 M}{B_0} = \mu_0 \frac{3N\mu_B^2}{2k_B T_F} \tag{6.61}$$

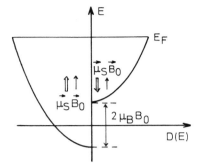

Fig. 6.15 Energy band shift in an external magnetic field B_0 (for free electron gas).

In contrast to the susceptibility of bound electrons with $j = 1/2$ ($l = 0$, $s = 1/2$), for which $\chi = \mu_0 N \mu_B^2/(k_B T)$ (for $\mu_B B \ll k_B T$), χ_{Pauli} does not depend on temperature. Because $T_F/T \approx 10^2$ at 300 K, the value of the Pauli susceptibility is approximately two orders of magnitude smaller than the susceptibility of bound electrons.

The magnetization depends on the spin polarization in an external field in the following way

$$M = N\gamma_e \langle S_z \rangle \tag{6.62}$$

Using Eq. (6.61), one obtains

$$\langle S_z \rangle = \frac{\chi_{\text{Pauli}}}{N\gamma_e} \frac{B_0}{\mu_0} = \frac{3}{2} \frac{\mu_B^2 B_0}{\gamma_e k_B T_F} \tag{6.63}$$

Next we shall find the magnetic field B_S produced at the nucleus by an electron spin S. The Hamiltonian operator describing the magnetic hyperfine interaction is

$$\mathcal{H}_{\text{hf}} = -\gamma_N (\mathbf{I} \cdot \mathbf{B}_S) \tag{6.64}$$

In the calculation of B_S, one must distinguish between electrons *outside* and *inside* the nuclear volume, i.e. one has to consider the dipole–dipole interaction of the former and the Fermi contact interaction for the latter.

Dipole–dipole interaction

A magnetic moment $\boldsymbol{\mu}_S = \gamma_e \mathbf{S}$ outside the nuclear volume can interact with the nuclear moment only by the dipole field. The dipole field at a distance r is

$$B_S(l \neq 0) = \frac{\mu_0}{4\pi}\gamma_e \frac{3(S \cdot r)r - Sr^2}{r^5} \quad (6.65)$$

Averaging over all directions yields $B_S = 0$ for s electrons. Eq. (6.65) thus gives nonzero contributions only for $l \neq 0$.

Fermi contact interaction

The s electrons possess a nonvanishing probability inside the nuclear volume and therefore can interact directly with the nuclear moment. For the derivation of this contribution we consider the magnetization $M(S)$ associated with the electron spin S

$$M(S) = \gamma_e S |\psi(r)|^2 \quad (6.66)$$

where $|\psi(r)|^2$ is the density of electrons at r. The magnetic field at the nuclear site in a magnetized medium (polarized electron cloud) is

$$B_S = \mu_0 M(S) - \frac{1}{3}\mu_0 M(S) = \frac{2}{3}\mu_0 M(S) \quad (6.67)$$

Here $(1/3)\mu_0 M(S)$ is the demagnetizing contribution of a sphere ($B_{\text{demag}} = -(1/3)\mu_0 M(S)$) which must be subtracted. Substitution of Eq. (6.66) in Eq. (6.67) gives for $r = 0$

$$B_S(l=0) = \frac{2}{3}\mu_0 \gamma_e |\psi(0)|^2 S \quad (6.68)$$

For the Hamiltonian operator of the hyperfine interaction one obtains

$$\mathcal{H}_{\text{hf}} = -\gamma_N (I \cdot B_S)$$
$$= -\frac{\mu_0}{4\pi}\gamma_N \gamma_e \left[\frac{3(S \cdot r)(r \cdot I) - (S \cdot I)r^2}{r^5} + \frac{8\pi}{3}|\psi(0)|^2(S \cdot I)\right] \quad (6.69)$$

The first term in Eq. (6.69) is the dipole–dipole interaction for $l \neq 0$, and the second term is the Fermi contact interaction for electrons with $l = 0$. A common abbreviated notation for the contact term is

$$a = \frac{\mu_0}{4\pi}\frac{8\pi}{3}\gamma_N \gamma_e \hbar^2 |\psi(0)|^2 \quad (6.70)$$

a is called the hyperfine interaction constant. The shift of the resonance frequency caused by the interaction described by Eq. (6.69) is called the Knight shift; in general the contact term gives the major contribution. In a strong magnetic field ($|\gamma_e S_z B_0| \gg |\mathcal{H}_{\text{hf}}|$), only the z component of S is nonzero, and we obtain for s electrons with $\Delta B = B_S$ ($l = 0$) from Eq. (6.68),

$$\frac{\Delta\omega}{\omega} = \frac{\Delta B}{B_0} = \frac{\mu_0}{4\pi}\frac{8\pi}{3}\gamma_e|\psi(0)|^2\frac{\langle S_z\rangle}{B_0} \tag{6.71}$$

Substituting $\langle S_z \rangle$ from Eq. (6.63) gives

$$K = \frac{\Delta\omega}{\omega} = \frac{2}{3}\chi_{\text{Pauli}}\frac{|\psi(0)|^2}{N} \tag{6.72}$$

The quantity K is called the Knight shift. Apart from a factor 2/3, K is equal to the Pauli susceptibility multiplied by an amplification factor $|\psi(0)|^2/N$, which gives the ratio of the electron density at the nucleus to the average electron density. K and χ_{Pauli} can be measured independently. Thus one can derive the enhancement (or diminution) of the electron density at the nucleus.

The term $|\psi(0)|^2$ in Eq. (6.72) can be expressed by the hyperfine constant a (Eq. (6.70)) which gives the relation

$$K = \frac{a\chi_{\text{Pauli}}}{\mu_0 N \gamma_N \gamma_e \hbar^2} \tag{6.73}$$

Some values of the Knight shift are summarized in Table 6.2. They can be negative as well as positive. One observes shifts up to several percent, but most are of order 0.1%. The Pauli susceptibility and the Knight shift are appreciably smaller than values obtained for bound paramagnetic electrons. The reason for that is that most conduction electrons are in states which are occupied by a spin-up and spin-down electron. Only a small fraction of electrons (those near the Fermi energy) contribute to the paramagnetism.

Table 6.2 Knight shift of some metals.

Metal	Li	Al	Cu	Pd	Pb
Knight shift (%)	0.0261	0.162	0.237	−3.0	1.47

6.6 Spin–Lattice Relaxation

The spin–lattice relaxation time T_1 was introduced phenomenologically in the Bloch equations. It is the characteristic time for the spin system to return to equilibrium after a B field is switched on. For this process to occur, it is necessary that the lattice and the spin system exchange energy quanta $\hbar\omega_{L,0}$, in order to establish a Boltzmann distribution of the M substates. The relaxation rate $1/T_1$ will thus depend on the probability that magnetic fluctuations at the Larmor frequency are present in the lattice system which then induce transitions in the nuclear spin system.

In the following we will limit the discussion to nuclei with spin $I = 1/2$, since in this case the considerations become very simple. We then have

$$\frac{1}{T_1} = W_\uparrow + W_\downarrow \approx 2W \tag{6.74}$$

$W_\uparrow$ is the probability per unit time for an excitation from a lower to a higher M level, and $W_\downarrow$ the probability per unit time for the opposite process. Since the population of the $+1/2$ and $-1/2$ M substates is only slightly different in NMR, both rates are equal to first order. In order to derive the spin–lattice relaxation time for a given physical system one must compute the rate W. This will be done for two important cases: (a) for diffusing nuclei which experience time-varying magnetic fields produced by the dipoles of neighboring atoms, and (b) for nuclei in metals in which conduction electrons produce fluctuating fields at the nucleus.

6.6.1 Spin–Lattice Relaxation by Atomic Motion

Here we will discuss the spin–lattice relaxation due to self-diffusion in a solid. The atom jumps from lattice site to lattice site by exchanging positions with a neighboring vacancy. In addition to the external B field, the nucleus experiences the sum of the B fields of the neighboring nuclei; the contribution of these dipolar fields varies from lattice site to lattice site, since we assume a statistical distribution of the nuclear spins over nuclear M substates. Thus the nucleus experiences a time-varying B field during diffusion. The mean residence time $\bar{\tau}$ of an atom at a lattice site can be described by the Arrhenius behavior

$$\frac{1}{\bar{\tau}} = \frac{1}{\tau_\infty} \exp\left(-\frac{Q}{k_B T}\right) \tag{6.75}$$

where τ_∞ is the mean residence time at infinite temperature. Its reciprocal value is the mean attempt frequency with which an atom strikes the barrier. $Q = E^F + E^M$ is the activation energy for self-diffusion, i.e. the sum of the energy E^F for the formation of a vacancy and the energy E^M for migration of a vacancy. The distribution of the residence times τ around the mean time $\bar{\tau}$ is usually approximated by a simple exponential, but it can be more complicated (see left side of Fig. 6.16)

$$P(\tau) = P_0 \exp\left(-\frac{|\tau|}{\bar{\tau}}\right) \tag{6.76}$$

The distribution of jump frequencies is obtained by Fourier transformation of $P(\tau)$

$$\begin{aligned} P(\omega) &= \frac{1}{2\pi} \int_{-\infty}^{\infty} P(\tau) \exp(-i\omega\tau)\, d\tau \\ &= \frac{2P_0}{\pi} \frac{\bar{\tau}}{1 + \omega^2 \bar{\tau}^2} \end{aligned} \tag{6.77}$$

The dipolar field experienced by the diffusing nucleus reflects this frequency distribution. The characteristic time for magnetic field changes is the correlation time τ_c which is, within a factor of order unity, the same as the mean residence time $\bar{\tau}$. The exact relation between τ_c and $\bar{\tau}$ must be calculated for each individual case. The fraction of the time-varying dipole field which oscillates with the Larmor frequency and is perpendicular to $\boldsymbol{B}_0$

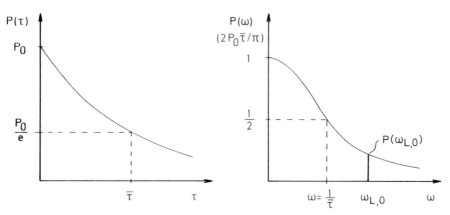

Fig. 6.16 Distribution of residence times between two jumps (left side) and corresponding distribution of jump frequencies (right side) for atomic self-diffusion.

produces transitions in the nuclear spin system. The transition rate W is thus proportional to the quantity $P(\omega = \omega_{L,0})$ (see right side of Fig. 6.16)

$$W \propto \frac{\tau_c}{1 + (\omega_{L,0})^2 \tau_c^2} \qquad (6.78)$$

Taking into account that the transition rate depends on the energy of the time-varying component of the magnetic field ($E \propto B_{dip}^2$), and assuming for simplicity that the field component varies only between the two values $+B_{dip}$ and $-B_{dip}$, one obtains for T_1 the following dependence

$$\frac{1}{T_1} \propto B_{dip}^2 \frac{\tau_c}{1 + (\omega_{L,0})^2 \tau_c^2} \qquad (6.79)$$

From this equation some important relations can be seen:

(a) The relaxation rate $1/T_1$ is strongest, or T_1 is smallest, for $\omega_{L,0} \tau_c = 1$, i.e. when the reciprocal of the correlation time is equal to the Larmor frequency.
(b) For $\omega_{L,0} \tau_c \gg 1$, i.e. when the nucleus performs many Larmor precessions during the mean residence time at a lattice site, one obtains ($\omega_{L,0}$ is fixed)

$$\frac{1}{T_1} \propto \frac{1}{\tau_c} \propto \exp\left(-\frac{Q}{k_B T}\right) \qquad (6.80)$$

(c) For $\omega_{L,0} \tau_c \ll 1$, i.e. when the nucleus performs many jumps during a Larmor precession, one obtains

$$\frac{1}{T_1} \propto \tau_c \propto \exp\left(+\frac{Q}{k_B T}\right) \qquad (6.81)$$

As can be seen, the relaxation rate is diminished by the rapid motion of the nucleus; that means that the different dipolar fields act so rapidly that they have little influence on the nucleus. This effect is also observable in the spin–spin relaxation time T_2 and in the linewidth $\Delta \omega = 2/T_2$ (see Eq. (6.28)) and is called motional narrowing.

Fig. 6.17 shows schematically the behavior of the relaxation rate $1/T_1$ as a function of the reciprocal temperature for two different Larmor frequencies, i.e. for two different external fields. An actual experimental example of this behavior will be treated in Section 6.7.

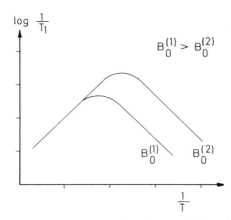

Fig. 6.17 Schematic relationship between the relaxation rate $1/T_1$ and reciprocal temperature for atomic self-diffusion for two different external magnetic fields B_0. The slopes of the curves are $+Q/k_B$ on the left of the maximum and $-Q/k_B$ on the right of the maximum (see Eqs. (6.80) and (6.81)).

6.6.2 Spin–Lattice Relaxation in Metals: Korringa Relation

Another important spin–lattice relaxation process occurs in metals. It is due to spin-flip scattering of conduction electrons by the nucleus. In this process the spins of the conduction electron and the nucleus are flipped. The interaction responsible for this process is the Fermi contact term (second term in Eq. (6.69)), i.e. the interaction which is also responsible for the Knight shift,

$$\mathcal{H}_{\text{hf}} = -\frac{2}{3}\mu_0 \gamma_N \gamma_e |\phi(0)|^2 (\mathbf{S} \cdot \mathbf{I}) \tag{6.82}$$

Only the diagonal terms contribute to the Knight shift, but the nondiagonal part of this interaction is responsible for the spin relaxation. The wave function $\phi(r)$ in Eq. (6.82) is normalized to one electron per unit volume.

Before discussing details we will first make a few general statements. The energy transfer in this scattering process, which involves only a flip of the electron and nuclear spins, is very small: $\Delta E \approx \hbar\omega_{L,0}(\text{nucleus}) - \omega_{L,0}(\text{electron})$; $\omega_{L,0}$ are the Larmor frequencies of the nuclear and electron spin respectively, in the external magnetic field. These energies are considerably smaller than the Fermi energy of conduction electrons and, at normal temperatures, much smaller than $k_B T$. This has the consequence that only electrons near the Fermi energy can contribute to the scattering process, since unoccupied levels are not accessible for other electrons. The total scattering probability depends on the product of the number of occupied and unoccupied levels. With the Fermi distribution of conduction electrons,

$$f(E) = \frac{1}{1 + \exp[(E - E_F)/k_B T]} \tag{6.83}$$

these considerations lead to the following relationship for the relaxation rate $1/T_1$,

$$\frac{1}{T_1} \propto f(E)[1 - f(E')] \tag{6.84}$$

The energies E and E' refer to the energies of the electron before and after scattering. It follows from the assumption $E \approx E'$ that

$$\frac{1}{T_1} \propto f(E)[1 - f(E')] = \frac{\exp[(E - E_F)/k_B T]}{\{1 + \exp[(E - E_F)/k_B T]\}^2}$$

$$= -k_B T \frac{df(E)}{dE} \tag{6.85}$$

or

$$\frac{1}{T_1} \propto f(E)[1 - f(E')] \approx k_B T \delta(E - E_F) \tag{6.86}$$

The relation (6.86) follows from (6.85) by integration of $df(E)/dE$ from 0 to ∞. The δ function takes into account that $f(E)[1 - f(E)]$ is different from zero only near $E = E_F$.

Eq. (6.86) shows that $1/T_1$ should be proportional to the temperature T. This proportionality results from the increase with T of the electrons participating in relaxation due to broadening of the Fermi edge.

For a quantitative derivation of the scattering formula we start with Fermi's golden rule for the transition probability,

$$W_{if} = \frac{2\pi}{\hbar} |\langle f|\mathcal{H}_{hf}|i\rangle|^2 \frac{dn_f}{dE_f} \tag{6.87}$$

Here again we will limit our discussion to nuclear spin 1/2. The generalization to higher spin poses no problems in general (see Abragam, 1961). In addition we will assume that we can work in the Paschen–Back region (high-field approximation). In this case the states can be denoted by

$$|i\rangle = |+, -\rangle \qquad |f\rangle = |-, +\rangle \tag{6.88}$$

Spin–Lattice Relaxation

Here $|+, -\rangle$ describes the state with $I_z = +\hbar/2$ (nucleus) and $S_z = -\hbar/2$ (electron) and $|-, +\rangle$ the opposite situation. The Hamiltonian operator in spherical coordinates has the form (compare Eq. (6.82))

$$\mathcal{H}_{\text{hf}} = -\frac{2}{3}\mu_0 \gamma_N \gamma_e |\phi(0)|^2 \left[I_z S_z + \frac{1}{2}(I_+ S_- + I_- S_+) \right] \quad (6.89)$$

From this expression we obtain for the transition probability (without the factor dn_f/dE_f)

$$W(|+, -\rangle \to |-, +\rangle) = \frac{2\pi}{\hbar} \left(\frac{2}{3}\mu_0 \gamma_N \gamma_e \right)^2 |\phi(0)|^4 \frac{\hbar^4}{4} \quad (6.90)$$

The total transition probability is obtained by multiplying $W(|+, -\rangle \to |-, +\rangle)$ by the statistical factor

$$D(E)f(E)D(E')[1 - f(E')] \quad (6.91)$$

where $D(E)f(E)$ is the number of scattering electrons and $D(E')[1 - f(E')]$ is the number of available final states. With Eq. (6.86) we obtain for the statistical factor

$$D(E_F)^2 k_B T \quad (6.92)$$

and finally for the relaxation rate $1/T_1$

$$\begin{aligned}\frac{1}{T_1} &= 2 \times \frac{2\pi}{\hbar} \left(\frac{2}{3}\mu_0 \gamma_N \gamma_e \right)^2 |\phi(0)|^4 \frac{\hbar^4}{4} D(E_F)^2 k_B T \\ &= \frac{4\pi}{9} \mu_0^2 \gamma_N^2 \gamma_e^2 \hbar^3 |\phi(0)|^4 D(E_F)^2 k_B T \end{aligned} \quad (6.93)$$

The factor 2 in Eq. (6.93) arises because spin flip is possible in both directions (see Eq. (6.74)). Eq. (6.93) is the final result of our consideration. For the free electron gas, $D(E_F)$ can be replaced by the expression in Eq. (6.57). However since $D(E)$ is now for one spin direction only, its value is a factor of 2 smaller than in Eq. (6.57). Substituting into Eq. (6.93) yields

$$\frac{1}{T_1} = \frac{\pi}{4} \mu_0^2 \gamma_N^2 \gamma_e^2 \hbar^3 |\psi(0)|^4 \frac{T}{T_F} \frac{1}{k_B T_F} \quad (6.94)$$

The total energy density at the nucleus has been introduced,

$$|\psi(0)|^2 = N|\phi(0)|^2 \quad (6.95)$$

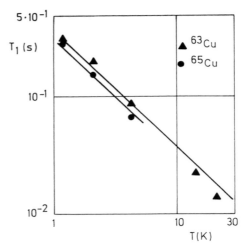

Fig. 6.18 Relaxation time T_1 for ^{63}Cu and ^{65}Cu in metallic copper as a function of temperature, shown as a log–log plot. The solid lines in the figure have slope -1 and therefore agree with the relation $T_1 \propto 1/T$ (Bloembergen, 1949).

Korringa relation

As one can easily verify, the Knight shift K in Eq. (6.72) and $1/T_1$ in Eq. (6.92) are related by

$$T_1 T = \frac{\hbar}{4\pi k_B}\left(\frac{\gamma_e}{\gamma_N}\right)^2 \frac{1}{K^2} \quad (6.96)$$

This is called the Korringa relation. In Eq. (6.96) one recognizes that $T_1 T$ (relaxation time multiplied by temperature) does not depend explicitly on temperature. This is indeed observed in many cases. An example of such a measurement is shown in Fig. 6.18.

6.7 NMR with Radioactive Nuclei and Self-Diffusion in Metals

Measurements with conventional NMR rely on the very small nuclear spin polarization caused by the Boltzmann distribution of the nuclear M states (see Section 6.1). In order to obtain a measurable induction signal, an appreciable number of probe nuclei is necessary (typically 10^{18} or more). However, if one uses radioactive nuclei whose spin has been polarized or aligned by a nuclear reaction, one can do an NMR experiment with many fewer nuclei (as few as 10^6). The unequal population of nuclear M substates is now independent of the applied B_0 field. The unequal population shows up as an anisotropic angular distribution of the emitted β or γ-radiation with respect to the beam axis. Experimentally one observes the NMR through the destruction of the anisotropic angular radiation distribution.

To obtain the polarization of radioactive nuclei one may use nuclear reactions with polarized proton or deuteron beams or the capture of polarized thermal neutrons. The resulting nuclear spin polarization can be detected through the β-decay of the nuclei, since the emission of the electrons or positrons is anisotropic with respect to the polarization axis of the nuclei.

Fig. 6.19 shows schematically the experimental set-up for NMR experiments with nuclei that decay by β-emission (β-NMR). In this case the nuclei are produced by polarized thermal neutrons. The neutron beam is pulsed by a chopper. Between the pulses the decay electrons (or positrons) are detected. Because of the nuclear spin polarization, one observes in the two β detectors a (north/south) asymmetry; this asymmetry can be destroyed by RF irradiation.

In case of nuclei that decay by γ-emission (γ-NMR), the nuclei are aligned by the angular momentum transfer in the nuclear reaction, and the anisotropic γ-distribution can be used as an indicator of the nuclear resonance (Quitmann et al., 1969). This variant of NMR will not be further treated here; it is presently used primarily to measure relaxation phenomena in liquid metals (Riegel et al., 1972).

The use of radioactive nuclei for NMR is useful only for nuclei with lifetimes in an appropriate range. As discussed in Section 6.2, the nuclear spins precess in resonance around the B_1 field, whereby the nuclear spin polarization is destroyed. If the nuclear state lives only a short time compared to the Larmor period $T(B_1) = 2\pi/\omega_{L,1}$ the polarization could not

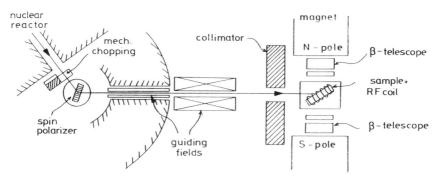

Fig. 6.19 Schematic diagram of a β-NMR experiment on nuclei produced by the capture of polarized thermal neutrons. Adapted from Winnacker et al. (1971).

be turned out of the z direction and therefore no nuclear resonance could be observed. Thus the following relation must be fulfilled

$$|\omega_{L,1}|\tau_N \gg 1 \quad \text{i.e.} \quad \gamma B_1 \tau_N \gg 1 \tag{6.97}$$

Since the B_1 field is limited for technical reasons ($B_1 \leq 2\,\text{mT}$), one obtains a lower limit for the nuclear lifetime. In general, the lifetimes of radioactive NMR nuclei are between 0.1 ms and 10 s.

As an example of a β-NMR experiment, the self-diffusion of ^{8}Li nuclei ($t_{1/2} = 0.8\,\text{s}$) in solid lithium metal will be discussed. Natural lithium metal consists of 92.5% ^{7}Li and 7.5% ^{6}Li. ^{8}Li is produced by neutron capture by ^{7}Li. In this experiment, the spin–lattice relaxation time was measured by observing the decay of the β-asymmetry ($A \propto \exp(-t/T_1)$) as a function of time following the production of the ^{8}Li in the neutron activation pulse.

The Korringa relaxation rate must be subtracted from the observed relaxation rate in lithium metal to obtain the spin–lattice contribution from self-diffusion. The result for such a diffusive relaxation rate $1/T_{1,\text{diff}}$ is shown in Fig. 6.20.

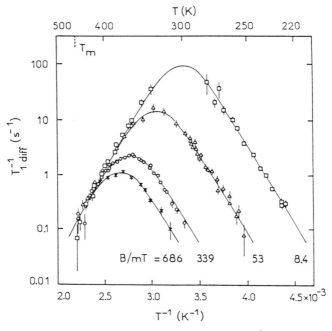

Fig. 6.20 Spin-lattice relaxation rate caused by self-diffusion of ^{8}Li in lithium metal as a function of reciprocal temperature for different external B_0 fields (Heitjans et al., 1985).

The dependence of the spin–lattice relaxation rate $1/T_1$ on temperature for diffusive motion was discussed extensively in Section 6.6.1. The relevant parameters can be deduced from the experimental results shown in Fig. 6.20. The self-diffusion energy Q in lithium metal can be determined from the slope of the measured curve to the right and left of the maximum (see Fig. 6.17):

$$Q = 0.57(2) \text{ eV}$$

Using the relation $\omega_{L,0}\tau_c = 1$, the position of the $1/T_1$ maximum as a function of temperature provides information about the absolute value of τ_c and therefore $\bar{\tau}$. The experiment whose results are shown in Fig. 6.18 provides the lithium self-diffusion rate in the temperature range between 220 K and the melting point ($T_m = 454$ K). The atomic residence times decrease from 10^{-3} to 10^{-9} s in this range.

7
Nuclear Orientation (NO)

7.1 Principles

Like NMR, the method of nuclear orientation (NO) uses the spin polarization of nuclei in thermal equilibrium in a B field. In contrast to NMR, nuclear orientation experiments are done at very low temperatures and very high B fields so that large nuclear spin polarizations are attained. The important quantity is the ratio of the level splitting ΔE to the thermal energy $k_B T$. In the case of magnetic interactions $\Delta E = E_{\text{magn}}(M+1) - E_{\text{magn}}(M) = -g\mu_N B_z$ (Eq. (3.3)). If

$$\Delta E / k_B T \gg 1 \tag{7.1}$$

is valid, then in thermal equilibrium the lower-lying M levels are populated much more strongly than the upper levels. If the level splitting is caused by an electric quadrupole interaction, only nuclear spin alignment occurs because of the M^2 degeneracy. Both, nuclear spin polarization and nuclear spin alignment are called nuclear orientation.

Nuclear orientation measurement is usually done with radioactive nuclei, since detection of the decay radiation anisotropy provides very high sensitivity. A famous example of nuclear orientation is the Wu experiment where ^{60}Co nuclei (see decay scheme in Fig. 5.3) were used to demonstrate parity violation in β-decay (Wu et al., 1957).

The physically important quantities for nuclear orientation will be discussed now. In order to reach the strong magnetic fields necessary to measure nuclear orientation, one often incorporates the radioactive nuclei into ferromagnetic substances. A small external B field of order 0.1 T is applied in order to align Weiss domains. For the nucleus ^{54}Mn often used in experiments, the internal B field in nickel is $B_{\text{loc}} = -32.55$ T. The resulting splitting of the 3^+ ground state with $\mu_g = +3.28\,\mu_N$ is displayed in Fig. 7.1. The occupation probabilities $P(M)$ in thermal equilibrium (Boltzmann

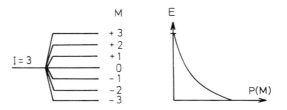

Fig. 7.1 Magnetic field splitting of the ground state of ^{54}Mn in nickel. The occupation probability is shown schematically on the right side of the figure.

distribution) is also shown,

$$P(M) \propto \exp[-E_{\text{magn}}(M)/k_B T] \tag{7.2}$$

The temperature at which the ratio $\Delta E/k_B T$ becomes unity for ^{54}Mn in nickel is

$$T = \frac{\Delta E}{k_B} = \frac{(\mu_g/I_g)B_{\text{loc}}}{k_B} = 13.1 \text{ mK} \tag{7.3}$$

At 10 mK the ratio of the occupation probabilities of the two extreme M states is

$$P(+3) : P(-3) = 1 : 2592 \tag{7.4}$$

Thus only the lowest state is occupied significantly at very low temperature, and the nuclear polarization is optimal.

Anisotropy of γ-radiation

If a polarized nuclear system decays by γ-radiation, the angular distribution is not isotropic in general. This situation was discussed in Chapter 5, so that only the final results for the distribution of γ-rays (see Eq. (5.19)) is given here,

$$W(T, \theta) = \sum_{k \text{ even}}^{k_{\text{max}}} \rho_k(I, T) U_k A_k(2) P_k(\cos \theta) \tag{7.5}$$

Here T is the temperature, and θ is the angle between the γ-emission direction and the axis of the nuclear orientation. The statistical tensor $\rho_k(I, T)$ (see Eq. (5.20)) contains the occupation probability $P(M)$ of the initial state. This probability depends on temperature via the Boltzmann distribution (Eq. (7.2)). One obtains

$$p_k(I, T) = \sqrt{(2k + 1)(2I + 1)} \sum_M (-)^{I-M} \begin{pmatrix} I & I & k \\ M & -M & 0 \end{pmatrix} P(M) \quad (7.6)$$

In the decay of an oriented nuclear spin system an intermediate step, e.g. β-decay or electron capture, is often not observed. In this case the 'de-orientation' parameters U_k due to the unobserved intermediate transitions are introduced. The quantities $A_k(2)$ are tabulated in Ferguson (1965) and $P_k(\cos\theta)$ are the Legendre polynomials.

For example the count rate of γ-radiation from the decay of ^{54}Mn measured at $\theta = 0°$ (Lounasmaa, 1974) is

$$W(T, 0°) = 1 + 0.38887 \sum_M M^2 P(M) - 0.055553 \sum_M M^4 P(M) \quad (7.7)$$

This expression contains only the temperature-dependent occupation probabilities $P(M)$ which can be measured by NO. Knowing all correlation parameters and occupation probabilities $P(M)$, the temperature of the nuclear system can be derived. Thus nuclear orientation can be used as a thermometer for very low temperatures.

7.2 Experimental Apparatus

The experimental set-up is relatively simple. One needs a NaI or Ge detector with which the emitted γ-radiation at a fixed angle (e.g. $\theta = 0°$) is detected. The normalization of the count rate is obtained from a measurement at a higher temperature where no polarization is present. The measurable quantity $W(T, 0°)$ is obtained via

$$W(T, 0°) = \frac{N_c}{N_w} \quad (7.8)$$

where N_c and N_w are the count rates for cold and warm samples. The most important part of the experimental set-up is the cryostat with which the low temperatures are attained. Nowadays a continuous ^{3}He–^{4}He dilution refrigerator is used.

7.2.1 ^{3}He–^{4}He Dilution Refrigerator

The principle of the ^{3}He–^{4}He dilution refrigerator is based on a continuous 'evaporation' of liquid ^{3}He into liquid ^{4}He. In this process heat is absorbed from the surroundings. ^{3}He is removed from the ^{4}He bath by pumping at another position in the system. It is important that ^{3}He is soluble in ^{4}He to a certain extent even at the lowest temperatures.

The ^{3}He–^{4}He phase diagram is shown in Fig. 7.2. Above the coexistence curve, ^{3}He and ^{4}He are completely miscible. On cooling such a mixture (e.g. 20% ^{3}He, 80% ^{4}He) below the coexistence curve, the two phases separate spontaneously: the system forms a ^{3}He-rich phase which because of the lower density floats on top, and a ^{4}He-rich phase which sinks to the bottom. Below 100 mK the ^{3}He-rich phase is almost pure ^{3}He in thermodynamic equilibrium whereas the ^{4}He-rich phase still contains (6.4% at $T = 0$) ^{3}He. We show later that it is possible to remove ^{3}He from the ^{4}He-rich bath by fractional distillation. It is then possible to 'evaporate' additional ^{3}He from the ^{3}He-rich phase into the ^{4}He-rich phase and thereby cool the mixture.

The principles of ^{3}He–^{4}He dilution refrigeration are illustrated in Fig. 7.3. The ^{3}He coming from outside is first condensed by contact with a ^{4}He bath ($T \approx 1.1$ K). Then it is cooled to as low a temperature as possible by the exhaust gas before it enters the mixing chamber and mixes with the ^{4}He. ^{3}He is removed from the ^{3}He–^{4}He mixture by pumping in the distillation chamber. The temperature in this region is kept at approximately 0.7 K. At this temperature ^{3}He evaporates easily and can be pumped off, whereas ^{4}He evaporates much more slowly (fractional distillation). The ^{3}He reaches the distillation chamber from the mixing chamber through osmotic pressure. The lowest temperatures achievable by this process are approximately 5 mK.

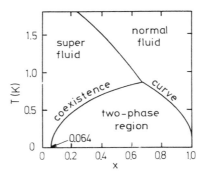

Fig. 7.2 Phase diagram of a ^{3}He–^{4}He mixture. $x = n(^3\text{He})/[n(^3\text{He}) + n(^4\text{He})]$ is the ^{3}He concentration. The tricritical point is at 0.84 K (Lounasmaa, 1974).

Experimental Apparatus   143

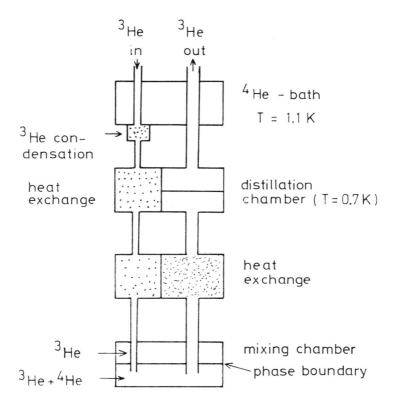

Fig. 7.3 Principal parts of a continuous dilution refrigerator.

7.2.2 *Radioactive Sources for Nuclear Orientation*

The nuclei ^{54}Mn and ^{177}Lu are examples of radioactive probe nuclei suitable for nuclear orientation.

^{54}Mn–^{54}Cr

The isotope ^{54}Mn can be produced by several different nuclear reactions. Starting with ^{51}V (99.75% isotopic abundance) as target, the nuclear reaction ^{51}V(α, n)^{54}Mn can be performed, but other reactions with proton beams are also possible, e.g. ^{54}Cr(p, n)^{54}Mn. Its decay is shown in Fig. 7.4.

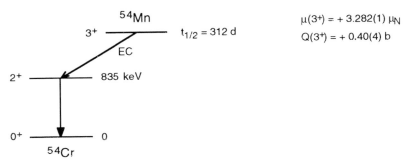

Fig. 7.4 Decay scheme of ^{54}Mn. The magnetic dipole moment and electric quadrupole moment of the 3+ probe state are given on the right side of the figure (Lederer and Shirley, 1978).

^{177}Lu–^{177}Hf

^{177}Lu can be produced most conveniently through neutron capture by ^{176}Lu (isotopic abundance 2.6%) in a reactor. The isotope ^{176}Lu in the 1$^-$ state with a halflife of 3.7 hours is also produced from ^{175}Lu (isotopic abundance 97.4%). This nucleus can also be used for nuclear orientation studies. Its decay is shown in Fig. 7.5.

Typical source strengths of the radioactive probes are several μCi (1 μCi = 3.7 × 10^4 decays/s). The source strength must be kept small in order to avoid warming the sample by absorption of the radiation. For ^{54}Mn the power absorbed (X-rays from the EC process) per μCi source strength is 50 pW. For β-emitters this power can be much larger. However, in order to

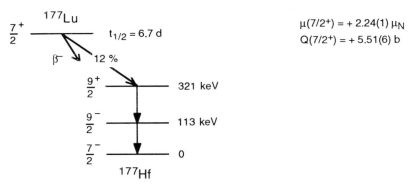

Fig. 7.5 Decay scheme of ^{177}Lu. The magnetic dipole moment and electric quadrupole moment of the 7/2+ probe state are shown on the right side of the figure (Lederer and Shirley, 1978).

reach the necessary count-rate statistics the source strength cannot be much lower than several μCi.

For long-lived ($t_{1/2} \geqslant$ several days) radioactive substances the probe can be mounted in the warm cryostat and then the cryostat can be cooled. A typical time to reach base temperature is one day. For short-lived radioactive substances, one must use a top-loading procedure, by which the probe is introduced from above into the already cold cryostat. However, the probe must be pre-cooled via contact at different stages in the cryostat. Such a procedure takes several hours. Very short-lived substances ($t_{1/2} \approx$ several minutes) can be introduced into the cold sample by implantation.

The spin–lattice relaxation imposes a lower limit on the lifetime of the isotope, since equilibrium between the nuclear spin system and the lattice must be reached before the measurement.

7.2.3 Nuclear Magnetic Resonance on Oriented Nuclei (NMR/NO)

Particularly high precision in determining the hyperfine splitting can be reached by a combination of nuclear magnetic resonance (NMR) and nuclear orientation (NO).

As we saw in the discussion of nuclear resonance (see Chapter 6) irradiation with RF frequency can remove the unequal occupation of the M substates, i.e. destroy the nuclear polarization. High sensitivity is reached by observing the resonance absorption via the disappearance of the anisotropic angular distribution. An example is shown in Fig. 7.6 for the case of ^{60}Co in ferromagnetic iron (Stone and Hamilton, 1984).

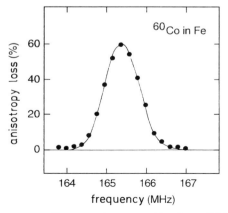

Fig. 7.6 NMR/NO resonance of ^{60}Co in Fe (Stone and Hamilton, 1984).

7.3 Hyperfine Fields

As we saw in the first section of this chapter, the anisotropy of the angular distribution depends, among other things, on the strength of the hyperfine splitting of the nuclear level. If all other parameters are known (e.g. via calibration on a known system), the hyperfine fields can be determined by measuring the anisotropy. Some of the internal B fields shown in Fig. 5.26 were measured by nuclear orientation.

An important point is that the sign of the hyperfine interaction can be determined by nuclear orientation under some circumstances. In the case of magnetic interactions, this is possible only if (parity-violating) β-radiation is measured, since in this case it makes a difference whether the $M = +I$ or $M = -I$ substate is the ground state. For γ-radiation, the sign of the ground M state is irrelevant, so the sign of the magnetic interaction cannot be determined.

For quadrupole splitting, the sign of the interaction can be determined even for γ-radiation. In this case either the large M values ($|M| = I$) or the small M values ($|M| = 0$ or $1/2$) form the ground state. An inversion of the term sequence leads to a change in the sign of the anisotropy. Thus for quadrupole interactions the sign of the hyperfine interaction can be obtained directly from the sign of the anisotropy. An example of pure quadrupole nuclear orientation is shown in Fig. 7.7.

7.4 Spin–Lattice Relaxation at Low Temperatures

If the temperature of the sample is changed in a nuclear orientation experiment, some time is required before the nuclear spin system reaches the new thermal equilibrium population. An example of such a measurement is shown in Fig. 7.8.

In this experiment the sample, in the mixing chamber of a ^{3}He–^{4}He dilution refrigerator, was induction-heated to a temperature of approximately 90 mK. After switching off the heating, the probe cools quickly (approximately 0.1 s) to the bath temperature of approximately 37 mK. The spin system follows this temperature change with a time delay.

Leveling-off of the anisotropy to a new constant value indicates that the new equilibrium substate population has been reached. The relaxation times in the present case are of order 10 to 100 s. These are typical times for metals at these temperatures. For paramagnetic probes in diamagnetic metals (e.g. ^{54}Mn in copper) these times may be much shorter (several μs) because of the amplification by the electron spin. On the other hand one observes very long relaxation times for semiconductors and insulators.

The dominant process for nuclear spin relaxation in nonmagnetic metals is

Spin–Lattice Relaxation at Low Temperatures   147

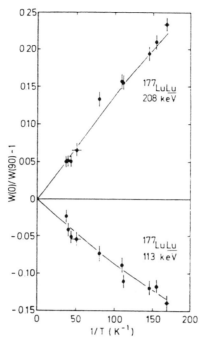

Fig. 7.7 Pure quadrupole nuclear orientation for ^{177}Lu in a Lu single crystal. By plotting the data vs. $1/T$, one obtains a straight line if the temperature is not too low (expansion of the exponential function $P(M)$) (Ernst *et al.*, 1979).

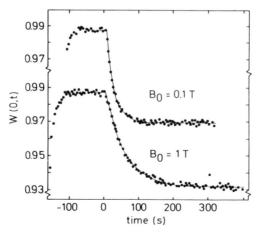

Fig. 7.8 Relaxation curves for ^{60}Co in iron at two different magnetic fields. For $B_0 = 0.1$ T the temperature was changed at $t = 0$ from 92 mK to 37.8 mK (upper curve). For $B_0 = 1$ T, the temperature was changed from 90 mK to 36.4 mK (lower curve). The spin system follows the cooling (new equilibrium M-level occupation) with a delay (Klein, 1977).

spin exchange scattering with conduction electrons (Korringa scattering). We discussed this process extensively in Section 6.6.2. An important assumption in deriving the formula in this Section was that the energy splitting of the nuclear levels is small compared to broadening of the Fermi edge. A consequence of this assumption was that the scattering probabilities from the higher to the lower level $W_\downarrow$ and from the lower to higher levels $W_\uparrow$ were equal. This assumption must be abandoned at the low temperature we consider here. The nonsymmetry is due to the fact that the process in which the nucleus goes from a higher to a lower level with simultaneous excitation of an electron near the Fermi energy to an unoccupied state is still possible, but the opposite process becomes increasingly less probable with decreasing temperature, since few electrons are available for this process.

Electrons in energy levels up to ΔE below the Fermi edge take part in the downward scattering $W_\downarrow$ in which electrons are scattered into free states above the Fermi edge. Electrons in lower-lying states cannot participate since they would be scattered into occupied states if they lost an energy ΔE. Thus we expect for $k_B T \ll \Delta E$

$$W_\downarrow \approx \Delta E \quad \text{and} \quad W_\uparrow = 0 \tag{7.9}$$

From the transition rates in Eq. (7.9) and the considerations discussed in Section 6.6.2 one can see that the relaxation rate becomes constant at low temperature. As can be seen in Fig. 7.9, this leveling-off at low temperature is observed experimentally.

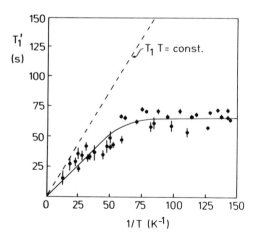

Fig. 7.9 Relaxation time T_1' for ^{60}Co in iron as a function of inverse temperature. The solid curve is a theoretical calculation. The dashed line corresponds to the Korringa relation. At low temperatures the relaxation rate levels off to a constant value (Brewer et al., 1968).

This low-temperature relaxation time is designated T'_1 to avoid confusion with the Korringa relaxation time and because different times occur for different sublevels at low temperature. Consequently T'_1 is an average over sublevels. An extensive theoretical treatment of this problem can be found in Bacon *et al.* (1972).

8
Muon Spin Rotation (μSR)

Muons belong to the lepton family. They exist in two charge states, positive (μ^+) and negative (μ^-), which are antiparticles of each other. The behavior of the two particles in solids is very different. The positive muon is comparable to a proton (H^+). It is repelled by atomic nuclei and generally resides in interstitial sites between the lattice atoms. The negative muon behaves in the solid like a heavy electron. It is attracted by the positive nuclei and is captured into Bohr orbits. Because of the much larger mass of the muon relative to the electron, these orbits are much closer to the nucleus than electronic orbits.

We will consider only the μ^+ spin rotation since positive muons are used much more extensively than negative muons in solid state physics research.

8.1 Principles

Production of muons occurs via pion decay. Pions must first be created in a high-energy proton collision ($E_p \geqslant 600$ MeV). Pions then decay with mean lifetime $\tau = 26$ ns into muons. The corresponding reactions are

$$p + p \rightarrow p + n + \pi^+$$
$$\pi^+ \xrightarrow{26 \text{ ns}} \mu^+ + \nu_\mu \tag{8.1}$$

It is important to note that muons produced in this way are 100% spin-polarized in the pion rest frame. Fig. 8.1 shows schematically the pion decay process. Muons and neutrinos both have spin 1/2, and the pion has spin 0. The neutrino is a massless particle and has helicity -1. From angular momentum conservation, the spin of the muon must therefore be directed antiparallel to its momentum.

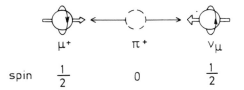

Fig. 8.1 Schematic diagram of π^+ decay. The spin of the muon is antiparallel to its emission direction. In the pion rest frame the energy of the emitted muon is 4.12 MeV.

The second important property of the muon in connection with μSR is the anisotropy of the muon decay,

$$\mu^+ \rightarrow e^+ + \nu_e + \bar{\nu}_\mu \tag{8.2}$$

Positrons (averaged over the e^+ spectrum) are emitted preferentially in the direction of the spin. Thus the muon spin direction can be inferred from measurement of the preferred positron emission direction.

The momentum distribution of positrons emitted in muon decay is shown in Fig. 8.2. The maximum energy is 52.83 MeV, and the average energy is

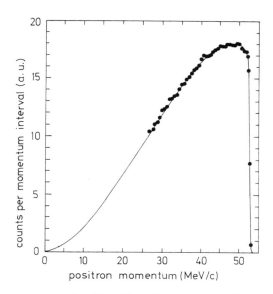

Fig. 8.2 Momentum spectrum of positrons from the decay of positive muons. Adapted from Bardon et al. (1965).

Table 8.1 Properties of the positive muon relevant to solid state physics.

spin	1/2
Mass	105.659 MeV/c^2 (206.769m_e)
Gyromagnetic ratio	8.5161 × 10^8 Rad s^{-1} T^{-1}
Decay	$\mu^+ \to e^+ + \nu_e + \bar{\nu}_\mu$
Mean (average) lifetime $\tau = t_{1/2} \ln 2$	2.197 × 10^{-6} s
Polarization in the rest system	100%
Angular distribution averaged over e$^+$ energies	$1 + 0.33 \cos \theta$
Character	light proton

36 MeV. Thus the positrons have relatively high energies and can be easily detected. Their range in aluminum is approximately 5 cm.

The most important properties of the muon are summarized in Table 8.1.

8.2 Experimental Methods

8.2.1 Muon Beams

'Arizona muons'

In the preceding section we saw that muons in the rest system of pions are 100% polarized. Thus if muons from *stopped* pions are used then one obtains a fully (antiparallel) polarized muon beam. Such a beam can be produced in practice. Since this was first done by an Arizona group, this beam is called an 'Arizona beam'. The primary proton beam (e.g. $E_p = 600$ MeV) produces pions in a beryllium target that is several centimeters long. Some of these pions come to rest in the production target and decay by emission of fully polarized muons with energy 4.1 MeV. These muons must be able to leave the target in order to be available for a beam. This is possible only if they are created near the surface of the target. Therefore these are also called 'surface muons'. The muons are formed into a muon beam and guided to the experimental set-up by a beam transport system that is several meters long and consists of quadrupole lenses and bending magnets. In spite of the restrictive conditions for creating Arizona muons, a muon beam of 10^6 to 10^7 μ^+/s can be obtained.

An essential advantage of Arizona muons is their complete polarization and the high stopping density (small range straggling). The latter property has the consequence that thin samples can be used. Because of the low energy of Arizona muons, they have the disadvantage that no windows, or only very thin windows, can be used in the beam transport line. The arrangement for producing Arizona muons is only suitable for positive muons. Because of their high capture cross section, negative Arizona muons

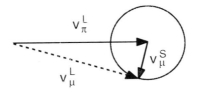

Fig. 8.3 Kinematics of the pion decay in flight. The velocity vectors of the pion and muon are shown in the rest frame (S) and the laboratory frame (L).

are largely captured in the beryllium target. Therefore no intense μ^- beam is possible in practice by this method.

Fast muons

Because of these disadvantages beams have been developed in which the pions decay *in flight*. In this case one first extracts pions (e.g. with a momentum of 220 MeV/c) from the beryllium target. These pions then decay in flight. Because of the different lifetimes of the pions (26 ns) and the muons (2.2 µs), at some distance down the beam, only muons will be present at an appreciable level. In order to obtain a high angular acceptance under optimal beam conditions, a superconducting solenoid (approximately eight meters long) is used. The pions and muons follow spiral paths in the magnet. In this way one obtains an intense muon beam. Because of the kinetic energy of the pions, muons produced in this way have a higher momentum (e.g. 120 MeV/c) and therefore can be guided to the sample more easily than slow Arizona muons. These beams have the disadvantages of lower stopping density and incomplete polarization. Incomplete polarization arises because muons in the beam include some emitted at nonzero angles with respect to the pion beam direction (see Fig. 8.3).

8.2.2 Experimental Apparatus

Fig. 8.4 shows schematically the angular distribution of the emitted positrons for a given spin direction of the muon. If the spin precesses in a magnetic field, then the emission probability also precesses, and one obtains a distribution of the following form

$$W(\phi, t) = 1 + A \cos(\phi - \omega_L t) \tag{8.3}$$

where ϕ is the angle between the polarization at time $t = 0$ and the observation direction; ω_L is the Larmor frequency. Averaging over the whole positron spectrum gives $A = 1/3$. It can be seen in Fig. 8.4 and Eq.

Fig. 8.4 Angular distribution of the emitted positrons from muon decay in a polar diagram. In the illustration the spin has rotated through the angle $\alpha = \omega_L t$ away from its original direction (antiparallel to the muon momentum).

(8.3) that the count rate in a spatially fixed detector oscillates with the Larmor frequency.

The μSR set-up is shown schematically in Fig. 8.5. A muon creates a signal in the detector M (muon counter) and is then stopped in the sample. This signal is used to start a clock provided detector E_1 (the positron counter, which is known generically as an 'electron' counter in high-energy physics applications) has not responded simultaneously. If detector E_1 has also detected a simultaneous event, then this is a false signal (e.g. nonstopped muons or passing positrons). The decay of the muons is registered in the telescope ($E_1 \wedge E_2$). This signal stops the clock provided there is no simultaneous signal from the M counter (in this case it would be a passing μ^+ or e^+). The clock transforms the time between start and stop into an address and increments the channel content corresponding to this address (see Section 5.4.3).

The logic box in Fig. 8.5 prevents processing an event by providing a reset pulse if the following situations occur:

(a) A second M signal occurs within a fixed time (e.g. 6 μs) after the first M signal. In such a case, it is impossible to determine which muon emitted the positron, and therefore the event is not recorded.
(b) Within this time two positron signals are registered.

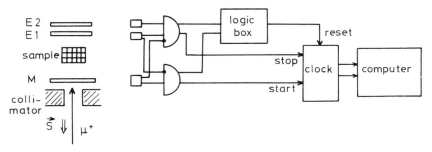

Fig. 8.5 Schematic diagram of μSR apparatus.

If more than one electron telescope is used, e.g. to increase the count rate, additional routing signals must be generated in order to sort the signals into different memory sectors. This is also done by the logic box.

The detectors consist of plastic scintillators (high time resolution) which are connected via light pipes to a photomultiplier tube. Light pipes must be used in order to reduce the magnetic field at the photomultiplier tube. The count rate of events in the E telescope as a function of time t after arrival of the muon in the sample is given by

$$N(t) = N_0 \exp(-t/\tau_\mu)[1 + P(t)A\cos(\phi - \omega_L t)] + B \qquad (8.4)$$

where τ_μ is the lifetime of the muon ($\tau_\mu = 2.2\ \mu s$), $P(t)$ the polarization as a function of time, and B a time-independent background caused by random events.

Fig. 8.6 shows schematically the μSR spectrum. One sees that the count rate decreases with the muon lifetime. An oscillation of period $2\pi/\omega_L$ is superimposed on this decay. In general more than one precession frequency ω_L^i can be present, namely when the muons experience different magnetic fields at different sites in the sample. One obtains the measurable quantities

$$P_i(t),\ A_i,\ \omega_L^i,\ \text{and}\ \phi_i \qquad (8.5)$$

$P_i(t)$ describes the behavior of the muon polarization at site i as a function of time, A_i is the portion of the total asymmetry due to muons precessing with frequency ω_L^i, and ϕ_i is the initial phase. In general ϕ_i is the angle between the initial polarization and the telescope direction. However, large interactions in precursor states at very short times may cause this angle to differ from the geometrical value.

The time differential measured spectrum $N(t)$ can be Fourier transformed to obtain the corresponding lines in frequency space (right side of Fig. 8.6).

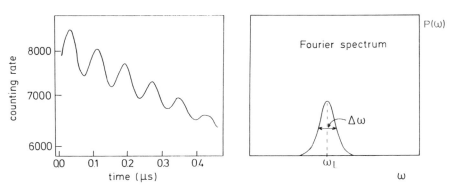

Fig. 8.6 Time differential and Fourier transform μSR spectrum (schematic).

The Fourier transform is particularly useful for indicating unknown frequencies which may not be easily recognized in the time spectrum. However, the time spectrum, not the Fourier transform, is normally fitted in order to determine the experimental parameters. In principle the time and Fourier transform spectra are physically equivalent.

8.3 Internal B Fields in Magnetic Materials

The muon is particularly well-suited as a probe of local magnetic fields in magnetic materials. The precession frequencies directly give the local B fields B_μ at the muon site via the relations

$$\nu_\mu = \frac{\gamma}{2\pi} B_\mu \quad \text{with} \quad \frac{\gamma}{2\pi} = 135.5 \, \text{MHz} \, \text{T}^{-1} \tag{8.6}$$

There are several interpretational difficulties. In general the muon site in the lattice is not known. One assumes that the muon in a good lattice is located in interstitial positions. In cubic crystals these are normally the particularly large tetrahedral and octahedral open spaces. Direct information about the muon site can be obtained by channeling experiments (Sigle et al., 1984).

Another difficulty is that the muon influences its surroundings, so its presence may alter the local magnetic field. This perturbation must be taken into account when comparing experimental results and theoretical predictions.

In ferromagnetic metals the local field B_μ can be separated in the following way (see Fig. 8.7 (Hellwege, 1976))

$$B_\mu = B_{\text{ext}} + B_{\text{dem}} + B_{\text{L}} + B_{\text{dip}} + B_{\text{Fermi}} \tag{8.7}$$

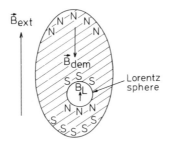

Fig. 8.7 Different contributions to the local magnetic field B_μ. N and S designate the north and south poles, respectively.

Here B_{ext} is the external field. B_{dem} is the demagnetization field due to the magnetic poles on the surface. B_{dem} depends on the form of the sample; for a sphere

$$B_{\text{dem}}(\text{sphere}) = -\frac{1}{3}\mu_0 M \tag{8.8}$$

where M is the macroscopic magnetization of the sample, and $\mu_0 = 1.256 \times 10^{-6}\,\text{V s A}^{-1}\text{m}^{-1}$ is the vacuum permeability. The Lorentz field B_{L} is the field at the center of a hypothetical spherical hole (Lorentz sphere). It has the value

$$B_{\text{L}} = \frac{1}{3}\mu_0 M_{\text{s}} \tag{8.9}$$

where M_{s} is the saturation magnetization. M_{s} can be determined in a macroscopic measurement. B_{dip} is the field due to individual dipoles within the Lorentz sphere and is given by (see Eq. (6.65))

$$B_{\text{dip}} = \frac{\mu_0}{4\pi}\sum_j \frac{3(\boldsymbol{\mu}_j \cdot \boldsymbol{r}_j)\boldsymbol{r}_j - \boldsymbol{\mu}_j r_j^2}{r_j^5} \tag{8.10}$$

Here $\boldsymbol{\mu}_j$ is the dipole moment of the lattice atoms and $\boldsymbol{r}_j$ is the position of the dipole with respect to the probe. Relaxation of atoms surrounding the muon should be taken into account. The hyperfine or Fermi contact field B_{Fermi} is caused by the polarization of conduction electrons. It was derived in section 6.5 and is given by

$$B_{\text{Fermi}} = -\frac{2\mu_0}{3}\mu_e \rho_{\text{spin}}(0) \tag{8.11}$$

$\rho_{\text{spin}}(0)$ is the electron spin density at the muon site.

Most μSR experiments on magnetic materials are performed without an external field and on demagnetized samples (e.g. by previously annealing the sample). In this case $B_{\text{ext}} = B_{\text{dem}} = 0$, and one obtains

$$B_\mu = B_{\text{L}} + B_{\text{dip}} + B_{\text{Fermi}} \tag{8.12}$$

Since B_{L} and B_{dip} (the latter with the restrictions mentioned above) can be calculated, the measured quantity B_μ gives directly the physically interesting Fermi contact field B_{Fermi}.

EXAMPLE μ^+ in nickel

Nickel has a face-centered cubic crystal structure. In this case the dipolar field at tetrahedral and octahedral interstitial sites is zero because of the cubic symmetry. In analogy to copper, which is also face-centered cubic and for which the muon site has been determined, one assumes that the muon in nickel occupies octahedral sites (see Fig. 8.8). With that assumption, Eq. (8.12) reduces to

$$\boldsymbol{B}_\mu = \boldsymbol{B}_\mathrm{L} + \boldsymbol{B}_\mathrm{Fermi} \tag{8.13}$$

In Fig. 8.9 the temperature dependence of the measured μSR frequency ν_μ and the magnitude of the B field derived from it are given. The temperature dependence of ν_μ and B_μ approximately follows the macroscopic magnetization curve (dashed line).

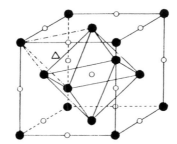

Fig. 8.8 Unit cell of the face-centered cubic (fcc) lattice. The full circles indicate the lattice positions; the open circles are the octahedral interstitial sites. In the upper front corner a tetrahedral site is indicated (open triangle).

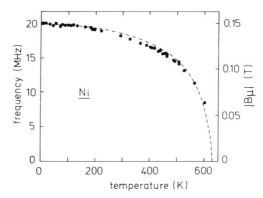

Fig. 8.9 Temperature dependence of the muon spin precession in nickel. The dashed line gives the temperature dependence of the magnetization normalized to the muon value at $T = 0\,\mathrm{K}$ (Denison *et al.*, 1979).

Extrapolation of the measured curve to $T = 0$ K gives

$$B_\mu = +0.149(1)\, \text{T} \quad \text{and} \quad B_{\text{Fermi}} = -0.072(1)\, \text{T} \quad (8.14)$$

Here B_μ is the measured quantity, and B_{Fermi} is derived from it by using Eq. (8.13) and the known value of the Lorentz field $B_L = +0.221$ T (Denison et al., 1979). The positive sign of B_μ was determined in an external magnetic field. In an external field one finds that the frequency ν_μ increases with increasing field after the sample magnetization reaches its saturation value. If ν_μ were negative, it would decrease. The value of B_{Fermi} agrees approximately with the value calculated for octahedral sites in an undisturbed lattice using the known magnetization. However, this agreement is somewhat accidental, since in general the spin density is strongly affected by the charge of the muon.

EXAMPLE μ^+ in iron

In the temperature range of interest ($T < 1180$ K), iron has a body-centered cubic structure. For this structure the tetrahedron surrounding the tetrahedral interstitial site and the octahedron surrounding the octahedral interstitial position (see Fig. 8.10) are distorted from ideal. As a consequence, the dipolar fields do not vanish at these sites. For further discussion we will assume that the muon is at a tetrahedral interstitial position. This is suggested by channeling experiments on hydrogen in bcc metals and pion experiments on the bcc metal Ta (Maier, 1984). It is important to realize that the dipolar fields at the tetrahedral interstitial positions are not the same but depend on the relative orientation of the tetragonal axis of

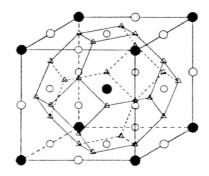

Fig. 8.10 The unit cell of a body-centered cubic (bcc) lattice. The lattice atoms are indicated by full circles. Tetrahedral (open triangles) and octahedral (open circles) interstitial sites are also shown.

the distorted tetrahedron with respect to the direction of the magnetization. Without an external field the magnetization of iron in a Weiss domain is along a cube edge direction (e.g. $\langle 001 \rangle$). The tetragonal axis can be either perpendicular or parallel to this axis. The dipolar fields for the tetrahedral sites calculated using Eq. (8.10) are

$$B_{\text{dip}}^{\parallel} = -0.52 \text{ T} \quad \text{and} \quad B_{\text{dip}}^{\perp} = +0.26 \text{ T} \quad (8.15)$$

where $B_{\text{dip}}^{\parallel}$ refers to parallel and $B_{\text{dip}}^{\perp}$ to the perpendicular orientation of the tetragonal axis with respect to the magnetization. Sites with $B_{\text{dip}}^{\perp}$ are twice as abundant as those with $B_{\text{dip}}^{\parallel}$. Since the sites are energetically indistinguishable, muons will occupy these sites statistically. Thus for a nondiffusing muon one would expect two different frequencies. However, only one frequency is observed experimentally. The implication is that the muon is diffusing rapidly even at low temperatures, therefore averaging out the dipolar fields. Thus on average, B_{dip} is zero in iron. Therefore the local field and the Fermi contact field can be calculated from the measured μSR frequency and the known Lorentz field $B_{\text{L}} = 0.73$ T (Denison et al., 1979). One obtains (extrapolated to $T = 0$ K)

$$B_\mu = -0.38(1) \text{ T} \quad \text{and} \quad B_{\text{Fermi}} = -1.11 \text{ T} \quad (8.16)$$

The Fermi contact field found in this experiment is a factor of 10 larger than the value obtained for an undisturbed lattice in neutron scattering experiments. Thus the muon enhances the spin polarization by a factor of 10. The temperature dependence of B_{Fermi} is shown in

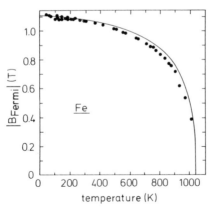

Fig. 8.11 Fermi contact field at the muon site in iron. The solid line gives the temperature dependence of the macroscopic magnetization (normalized to the experimental field at low temperatures) (Denison et al., 1979).

Fig. 8.11. One sees that the temperature dependence of B_{Fermi} qualitatively follows the macroscopic magnetization.

8.4 Diffusion of Positive Muons

The positive muon is at rest in the solid in only a few cases. It is usually diffusing rapidly by jumping among interstitial positions. The muon diffusion is particularly interesting, because a positive muon in the solid can be considered a light proton. This extends the range of isotopic mass (from the muon with mass $m_H/9$ to the triton with mass $3m_H$) available for experimental diffusion studies. Because of their small mass, muons are particularly useful for studying quantum effects, which are important at low temperatures.

Two different principles (motional narrowing and capture at defects) are used to measure the muon diffusion. In the following we first discuss these two principles and give examples of the application. In Section 8.4.3 we describe some diffusion models.

8.4.1 Motional Narrowing

In NMR, the linewidth often decreases at higher temperature due to particle motion (see Section 6.6.1 on motional narrowing). The same effect is observed in μSR. The direct analogy is narrowing of the linewidth in the Fourier spectrum. In the time spectrum this corresponds to a decrease of the relaxation rate.

Static muon

The starting point for a discussion of motional narrowing is the assumption of a distribution of local magnetic fields in the crystal. We will assume here that this distribution is caused by the nuclear moments of the atoms in the lattice near the muon. In diamagnetic metals the dipole–dipole interaction is the main source of locally varying magnetic fields.

If the local magnetic fields have a Gaussian distribution around the average value of the external field B_0, then the muon precession frequencies also have a Gaussian distribution

$$f(\omega) = \frac{1}{\sqrt{2\pi}\,\sigma} \exp\left[-\frac{(\omega - \omega_{L,0})^2}{2\sigma^2}\right] \qquad (8.17)$$

Using Eq. (8.17) as the weighting function yields

$$\overline{\Delta \omega^2} = \int_{-\infty}^{+\infty} (\omega - \omega_{L,0})^2 f(\omega) \, d\omega = \sigma^2 = \gamma_\mu^2 \overline{\Delta B^2} \qquad (8.18)$$

In the time spectrum one obtains a superposition of precession signals with different frequencies. Averaging over the muon ensemble gives the result

$$\overline{\cos \omega t} = \frac{1}{\sqrt{2\pi} \, \sigma} \int_{-\infty}^{+\infty} \exp\left[-\frac{(\omega - \omega_{L,0})^2}{2\sigma^2}\right] \cos \omega t \, d\omega$$

$$= \exp\left(-\frac{\sigma^2 t^2}{2}\right) \cos \omega_{L,0} t \qquad (8.19)$$

As can be seen in Eq. (8.19) the muon signal is centered at $\omega_{L,0}$ and damped with a Gaussian envelope.

The magnetic fields caused by nuclear dipoles are of order 10^{-4} T. Assuming this for the RMS value, one obtains from Eq. (8.18) σ values of order $0.1 \, \mu s^{-1}$. A rigorous description requires knowledge of the muon site in the lattice and the orientation of the external magnetic field. Under normal conditions (B_0 large with respect to fields produced by the muon magnetic moment at neighboring nuclei),

$$\sigma^2 = \gamma_\mu^2 \frac{1}{3} I(I+1) \hbar^2 \gamma_I^2 \left(\frac{\mu_0}{4\pi}\right)^2 \sum_j \frac{(1 - 3\cos^2 \theta_j)^2}{r_j^6} \qquad (8.20)$$

where γ_μ and γ_I are the gyromagnetic ratios of the muon and nucleus respectively, I the nuclear spin, r_j the position of the atom j with respect to the muon, and θ_j the angle between r_j and the external field B_0. B_{dip} in Eq. (8.20) has the same structure as B_{dip} in Eq. (8.10) except that only the z component of μ_j is retained because the strong magnetic field makes other components negligible. For a detailed derivation of Eq. (8.20) see Seeger (1978).

The strong dependence of σ^2 in Eq. (8.20) on the position of the nearest-neighbor atoms (r_j) and on the orientation of the external magnetic field (θ_j) can be used to determine the site of the muon in the crystal. Using this method in copper, it was found that the muon occupies octahedral sites and pushes the nearest-neighbor atoms outward by 5% (Camani et al., 1977).

For polycrystalline samples, the angular-dependent part in Eq. (8.20) must be averaged over all spatial directions. This yields

$$\overline{(1 - 3\cos^2 \theta_j)^2} = \frac{4}{5} \qquad (8.21)$$

Diffusing muon

If the muon diffuses, it averages over the different magnetic fields, and the damping of the muon spin precession decreases. The correlation time τ_c is an important quantity for the description of this effect. Qualitatively it is the time which the muon needs to diffuse into a region where the magnetic field is considerably different from the value at its original position. The exact definition is given in terms of the correlation function

$$g(t') = \langle B(t)B(t - t')\rangle_t \qquad (8.22)$$

where $\langle \ \rangle_t$ indicates time averaging. One defines τ_c as the time at which

$$g(\tau_c) = \frac{g(0)}{e} \qquad (8.23)$$

The motional narrowing becomes effective when

$$\sigma\tau_c \leq 1 \qquad (8.24)$$

since the magnetic field changes occur before the polarization has decreased significantly. Quantitative considerations (Seeger, 1978) which we will not discuss in detail give for the time dependence of the polarization

$$P(t) = P(0)\exp\{-\sigma^2\tau_c^2[\exp(-t/\tau_c) - 1 + t/\tau_c]\} \qquad (8.25)$$

Here σ is the depolarization rate of the static muon, and τ_c the correlation time defined above. For static muons ($\tau_c \to \infty$) Eq. (8.25) must of course become the same as Eq. (8.19). One sees that easily if one expands $\exp(-t/\tau_c)$ in Eq. (8.25) to second order.

The second extreme case occurs for rapidly diffusing muons. If

$$\sigma\tau_c \ll 1 \qquad (8.26)$$

then the important time region for $P(t)$ is given by $t \gg \tau_c$ so that in Eq. (8.25) the term t/τ_c dominates. This leads to the so-called motional narrowing formula

$$P(t) = P(0)\exp(-\lambda t) \qquad (8.27)$$

with

$$\lambda = \sigma^2\tau_c \qquad (8.28)$$

For very short times ($t \leq \tau_c$) the time dependence is not correctly described by Eq. (8.27); however, this is not experimentally important, since the very small-time behavior is not observable. In the intermediate region, in which $\sigma\tau_c \approx 1$, the full Eq. (8.25) must be used.

EXAMPLE μ^+ in copper

The classic example for μ^+ diffusion is the measurement by Grebinnik *et al.* (1975) in copper. In the μSR spectrum in Fig. 8.12 one can clearly see damping at low temperatures; at 330 K this damping is no longer visible.

As a measure of the damping, in Fig. 8.13, the rate t_e^{-1} is shown, where t_e is the time after which the polarization has decreased to the value $1/e$. In Fig. 8.13 the beginning of motional narrowing is recognizable at $T = 100$ K as a decrease of the depolarization rate. At lower temperatures a constant value of t_e^{-1} is observed. This value corresponds to the static value of σ and is

$$\frac{\sigma}{\sqrt{2}} = 0.266 \times 10^6 \text{ s}^{-1} \qquad (8.29)$$

It agrees well with the value calculated according to Eq. (8.20) for static muons in copper.

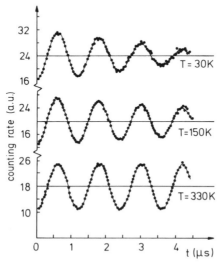

Fig. 8.12 The μSR spectra for copper at three different temperatures. The external magnetic field was 6.2 mT (Grebinnik *et al.*, 1975).

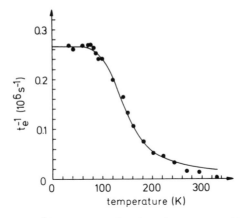

Fig. 8.13 Relaxation rate t_e^{-1} in copper as a function of temperature (Grebinnik et al., 1975).

If the value of σ is fixed, the value of τ_c can be determined by fitting the measured curves with Eq. (8.25) or with the simpler equations (8.27) and (8.28).

The diffusion coefficient can be derived from the correlation time τ_c. For that derivation we make the following assumptions, which are suggested by the experiment: (a) the muon diffuses between octahedral sites in copper and (b) the correlation time τ_c is approximately equal to the mean residence time $\bar{\tau}$ of the muon at a given site. The general diffusion theory then gives the following relation between the diffusion coefficient D and the correlation time τ_c

$$D = \frac{a^2}{12\tau_c} \tag{8.30}$$

This equation is valid for fcc lattices and diffusion between octahedral sites; a is the lattice constant.

8.4.2 Trapping by Lattice Defects

Lattice defects such as impurities or vacancies often have positive binding energies for muons and therefore can capture freely diffusing muons (Fig. 8.14). If one assumes that the muons are statistically distributed in the crystal after implantation, then the muon diffusion from the time of implantation to the arrival of the muon at a defect can be derived if the defect concentration is known. With several generally uncritical assumptions, the following equation holds

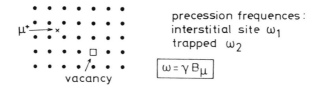
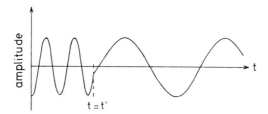

Fig. 8.14 Muon diffusion, starting from a random distribution of a muon in the lattice until it is trapped at a vacancy. In the lower part of the figure, the change of the precession frequency at the instant of trapping is indicated.

$$D = \frac{V_A}{4r_0 c_D} \frac{1}{\tau_D} \qquad (8.31)$$

where c_D is the defect concentration per atom, r_0 the capture radius (the distance within which the muon will be captured), V_A the atomic volume, and τ_D the time from the appearance of the muon in the sample until it is trapped at the defect. τ_D is determined in the μSR experiment.

EXAMPLE Defects in iron

Möslang *et al.* (1983) have irradiated an iron sample at low temperatures with electrons in order to produce vacancies in the sample. The concentration of vacancies (approximately 10^{-5}/atom) can be determined roughly by measuring the residual resistivity at 4 K. In the μSR experiment the muons are stopped initially in unperturbed regions of iron, since the concentration of defects is low. At this time ($t = 0$), the muon spin begins precessing in the local magnetic field of iron. The precession frequency in this case is 50 MHz. After precessing at this rate during free diffusion, the muons are trapped at vacancies where they occupy substitutional lattice sites having a different local magnetic field. This leads to a change in the precession frequency beginning at the time of trapping. Of course the trapping times have a statistical distribution. The frequency change is clearly seen in Fig. 8.15. From the time of the frequency change (curve t_D in Fig. 8.15) the

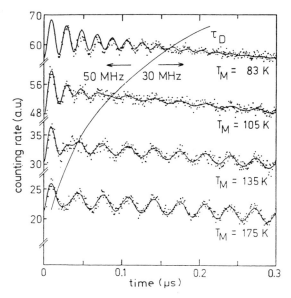

Fig. 8.15 The μSR signals of electron-irradiated iron. The precession frequency changes at approximately τ_D.

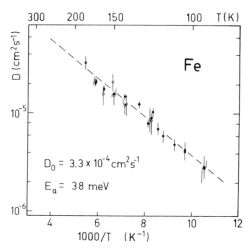

Fig. 8.16 Diffusion coefficients for muons in iron. Adapted from Möslang et al. (1983).

mean time τ_D required by the muon to come from an arbitrary stopping site to the vacancy can be determined (in practice the exact determination of τ_D is obtained from the amplitude of the second frequency). One sees in Fig. 8.15 that τ_D is strongly temperature-dependent as a result of temperature-dependent diffusivities.

The diffusion coefficient D can be derived from the measured value τ_D using Eq. (8.31). The result is shown in Fig. 8.16; r_0 is assumed to equal $3a$ (a is the lattice constant). The large uncertainties in r_0 and in the defect concentration c_D result in large systematic uncertainties for D. These uncertainties are not included in the error bars shown in Fig. 8.16.

8.4.3 Diffusion Models

Classic diffusion

The classic picture of the diffusion of an interstitial particle in a solid is shown schematically in Fig. 8.17. The particle must pass over a barrier of height E_a on its way from one potential minimum to the next. The probability that the particle possesses the energy required to pass over the barrier follows the Boltzmann distribution $f(E) \propto \exp(-E_a/k_B T)$. This gives, for the jump frequency v and the diffusion coefficient D, an Arrhenius behavior

$$v = v_0 \exp(-E_a/k_B T) \tag{8.32}$$

$$D = D_0 \exp(-E_a/k_B T) \tag{8.33}$$

v_0 is called the attempt frequency; its value is of order the Debye frequency ($v_0 \approx 10^{13}$ s^{-1}).

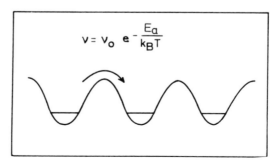

Fig. 8.17 Illustration of classic diffusion, in which the particle jumps over barrier E_a.

Incoherent tunneling

At low temperatures (room temperature and below in many cases) the jump frequency given by Eq. (8.32) is so small that little or no diffusion should occur. For hydrogen (e.g. in Nb and Ta) and, in particular, for muons the diffusion is often found not to freeze out but to remain observable down to very low temperatures. In these cases, diffusion takes place via a tunneling process (see Fig. 8.18).

In order to understand the temperature dependence of this process somewhat better, we must first consider the effect of self-trapping of a particle. Self-trapping occurs because the lattice reacts to the presence of the particle, e.g. by lattice expansion, thereby lowering the total energy. This is called a polaron. This implies that the occupied level is energetically lowered compared to the neighboring unoccupied level. For a tunneling process to occur, the two levels must first be brought to the same value through thermal fluctuations before the tunneling matrix element can become effective.

For all but extremely low temperatures the probability of equal energy values is given by the Arrhenius law,

$$W_1 \propto \exp(-E'_a/k_B T) \quad (8.34)$$

The activation energy is much smaller than the diffusional barrier energy (E_a) encountered in normal diffusion. The tunneling probability is given in perturbation theory by the square of the tunneling matrix element

$$W_2 \propto J^2 \quad (8.35)$$

If both probabilities are small, the total probability, which determines the jump rate v, is given by the product of these two individual probabilities

$$v \propto J^2 \exp(-E'_a/k_B T) \quad (8.36)$$

The diffusion properties are shown schematically in Fig. 8.19. It is assumed here that classic diffusion dominates at higher temperatures, so

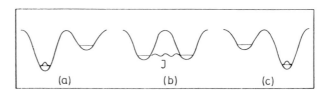

Fig. 8.18 Schematic representation of the incoherent tunneling process.

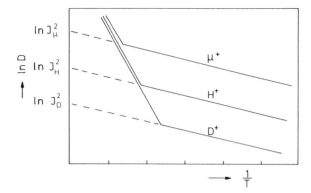

Fig. 8.19 Diffusion behavior of different 'hydrogen' isotopes (schematically). The kink indicates the transition from classic diffusion to incoherent tunneling.

only minor isotope dependences are expected. A kink in the Arrhenius lines should be observable at lower temperatures for all isotopes. However, since the tunneling matrix element is strongly mass-dependent, the kink should be shifted to lower temperatures with increasing mass. E'_a is expected to be mass-independent to first order, since the self-trapping is essentially due to electrical interaction which depends only on the charge of the particle.

As an experimental example the muon diffusion in copper (see Section 8.4.1) will be discussed. Fig. 8.20 shows the muon and hydrogen diffusion in copper in a common Arrhenius plot (Seeger, 1978). One recognizes clearly the much lower activation energy (corresponding to the slope of the Arrhenius plot) for muons than for hydrogen. It is implicitly assumed in this plot that at higher temperatures ($T > 300$ K) the muon diffusion and hydrogen diffusion coefficients are equal. At even lower temperatures, far below the region of measurement, one also expects a kink in the Arrhenius line for hydrogen. Such a kink has been observed experimentally for hydrogen in some bcc metals (Nb, Ta).

Coherent tunneling

In an ideal periodic potential the solutions of the Schrödinger equation for an interstitial particle are extended band states (Fig. 8.21), but the states can become localized if potential fluctuations are too large. The ratio of the bandwidth to the magnitude of the potential fluctuations plays a major role in determining whether states are extended or localized.

The bandwidth in an ideal periodic potential is given by the tunneling matrix element J, which can, in principle, be calculated for a given potential. However, the potentials in solids are usually not well known so

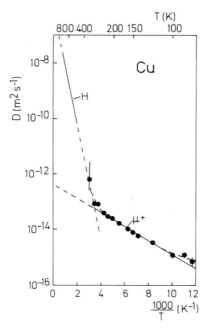

Fig. 8.20 Temperature dependence of the diffusion coefficients for hydrogen (steep line) and muon (data points) in copper (Seeger, 1978).

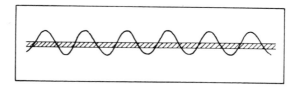

Fig. 8.21 Band states in an ideal periodic potential.

only rough estimates of J can be made. It is not just bare particles but also the lattice distortion and the electron shielding cloud that tunnel. We will call the resulting tunneling matrix element J_{eff}.

In a real crystal, deviations from the ideal potential occur because of impurities and lattice defects. The average shift ΔE between two neighboring potential minima can be taken as a measure of these deviations.

A necessary requirement for the appearance of extended band states is

$$J_{\text{eff}} > \Delta E \tag{8.37}$$

The opposite relation $\Delta E > J_{\text{eff}}$ is given in the literature as the condition for Anderson localization of electrons (Anderson, 1958). Because of Eq. (8.37) one expects extended band states for very pure and defect-free crystals at very low temperature.

The diffusion coefficient for diffusion in extended states is given by the following expression (Kehr, 1984)

$$D = \frac{1}{3}v^2 \tau_{\text{tr}} \qquad (8.38)$$

where v is the average velocity of the particle in the occupied band states, and τ_{tr} the transport lifetime, i.e. the time between two scattering processes. If τ_{tr} is mainly limited by the scattering from electrons, then the temperature dependence of D is given by (Kehr, 1984)

$$D \propto J_{\text{eff}}^2 T^{-1} \qquad (8.39)$$

It can be seen that for this process, D increases with decreasing temperature. Thus one expects diffusion coefficients at high temperature to be in the region of classic diffusion or incoherent tunneling and to decrease with decreasing temperature, but then below some temperature to increase again. An example of such behavior is observed for muon diffusion in aluminum (see Fig. 8.22).

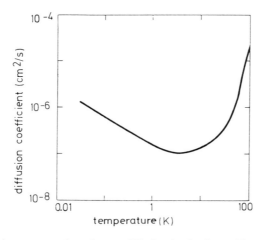

Fig. 8.22 Schematic representation of muon diffusion in aluminum. The values were obtained from an analysis of μSR data on Mn-doped aluminum using the trapping model (see Section 8.4.2) (Kehr, 1984).

8.5 Muonium in Semiconductors

In semiconductors and insulators the positive muon can capture an electron and form muonium (μ^+e^-). This state corresponds to non-ionized hydrogen in the solid. Owing to the unpaired electron, muonium is paramagnetic.

In metals such a state is not possible, since the electron is, in a sense, transferred to the conduction band. It is also correct to say that the electron at the muon is permanently exchanged with electrons of different spin directions, with the consequence that the behavior is diamagnetic. In the μSR experiment, such a state cannot be distinguished from a bare μ^+.

Muonium states have been detected in a number of insulator and semiconductors directly. In particular, there have been extensive investigations of SiO_2 and the pure semiconductors Si, Ge, and diamond. As examples we consider silicon and germanium in the following.

As will become clear later, different muon states can easily be distinguished via their hyperfine interaction. In Si and Ge three different states are found at low temperatures; normal muonium with a strong hyperfine interaction, anomalous muonium with a relatively weak hyperfine interaction (Patterson et al., 1978), and bare or diamagnetic μ^+. The two muonium states are present in silicon with approximately equal intensities. In the following we will discuss in more detail normal muonium and its behavior in external B fields.

8.5.1 Normal Muonium (μ^+e^-)

The spin part of the Hamiltonian in an external field B is

$$\mathcal{H} = \frac{a}{\hbar^2} J \cdot S - \gamma_\mu S \cdot B - \gamma_e J \cdot B \tag{8.40}$$

The last two terms describe the Zeeman energy of the muon spin S and the electron spin J in an external magnetic field respectively. The first term is the hyperfine interaction (Fermi contact interaction). Here we have assumed an isotropic interaction, which adequately describes normal muonium. We consider the strength of the hyperfine interaction a to be a free parameter; a depends via (see Eq. (6.70))

$$a = \frac{2}{3}\mu_0 \gamma_\mu \gamma_e \hbar^2 |\psi(0)|^2 \tag{8.41}$$

on the occupation probability $|\psi(0)|^2$ of the bound electron at the muon site. Except for substitution of γ_μ for γ_p, Eq. (8.40) is identical to the Hamiltonian operator of the 1s ground state of hydrogen. The energy

Muonium in Semiconductors 175

eigenvalues yield the known Breit-Rabi diagram. We will discuss here the two special cases for strong and weak magnetic fields in some detail.

8.5.2 Zeeman Region (Weak Magnetic Field)

In a weak magnetic field, the Zeeman terms of Eq. (8.40) are small compared to the hyperfine interaction energy a, so that the coupling scheme of Fig. 8.23 is valid.

In this case J and S precess rapidly around the total angular momentum

$$F = J + S \qquad (8.42)$$

whereas F precesses as a whole around the external field B. In this case, F and M_F as well as J and S are good quantum numbers but not M_J or M_S. After squaring Eq. (8.42) one obtains the hyperfine interaction energy

$$\frac{a}{\hbar^2}\langle J \cdot S \rangle = \frac{a}{\hbar^2}\left\langle \frac{1}{2}(|F|^2 - |J|^2 - |S|^2) \right\rangle$$

$$= \frac{a}{2}[F(F+1) - J(J+1) - S(S+1)]$$

$$= \frac{a}{4}[2F(F+1) - 3] \qquad (8.43)$$

where $\langle\ \rangle$ designates the quantum mechanical expectation value, and $J = S = 1/2$ has been used.

In order to determine the Zeeman energies, we first calculate the

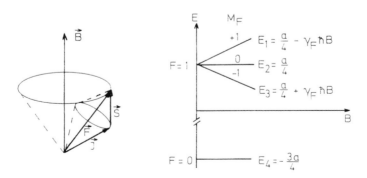

Fig. 8.23 Coupling scheme for J and S to F, and energy diagram for weak magnetic field B.

gyromagnetic ratio for F according to the generalized Landé formula (Eq. (2.11)) and obtain

$$\gamma_F = \frac{1}{2}(\gamma_\mu + \gamma_e) \approx \frac{1}{2}\gamma_e \quad \text{(since } |\gamma_\mu| \ll |\gamma_e|\text{)} \tag{8.44}$$

From this calculation we obtain for the Zeeman energy of F

$$-\gamma_F \langle \mathbf{F} \cdot \mathbf{B} \rangle = -\gamma_F B \hbar M_F \tag{8.45}$$

and finally

$$\langle \mathcal{H} \rangle = \frac{a}{4}[2F(F+1) - 3] - \gamma_F B \hbar M_F \tag{8.46}$$

with $F = 0$ or 1 and $|M_F| \leq F$. The resulting level scheme is displayed in Fig. 8.23 (note γ_e is negative!).

8.5.3 Paschen–Back Region (Strong Magnetic Field)

If one of the Zeeman terms in Eq. (8.40) is large compared to the interaction term then $\mathbf{J}$ and $\mathbf{S}$ are decoupled, and each of these angular momenta precesses separately around the external magnetic field (compare Fig. 8.24).

In this case J and M_J, S and M_S are good quantum numbers. In first-order perturbation theory, one obtains

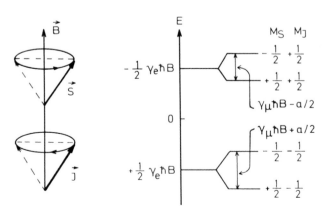

Fig. 8.24 Coupling scheme and energy levels in strong magnetic fields.

$$\langle \mathcal{H} \rangle = \frac{a}{\hbar^2} \langle J_z S_z \rangle - \gamma_\mu \langle S_z \rangle B - \gamma_e \langle J_z \rangle B$$
$$= a M_J M_S - \gamma_\mu B \hbar M_S - \gamma_e B \hbar M_J \qquad (8.47)$$

with $M_J = \pm 1/2$ and $M_S = \pm 1/2$. The resulting level scheme is displayed in Fig. 8.24. We note that $|\gamma_\mu| \ll |\gamma_e|$ and that γ_e is negative.

8.5.4 General Case

The general solution of the Hamiltonian operator, Eq. (8.40), yields the energy eigenvalues (Celio and Meier, 1983)

$$E_1(M_F = +1) = \frac{a}{4} - \frac{\gamma_e + \gamma_\mu}{2} \hbar B$$
$$E_3(M_F = -1) = \frac{a}{4} + \frac{\gamma_e + \gamma_\mu}{2} \hbar B \qquad (8.48)$$
$$E_{2,4}(M_F = 0) = -\frac{a}{4} \pm \frac{a}{2} \sqrt{1 + x^2}$$

with

$$x = -\frac{\gamma_e - \gamma_\mu}{a} \hbar B \qquad (8.49)$$

The eigenstates have the following form

$$\begin{aligned}\psi_1 &= |+\rangle_\mu |+\rangle_e \\ \psi_3 &= |-\rangle_\mu |-\rangle_e \\ \psi_2 &= \alpha |+\rangle_\mu |-\rangle_e + \beta |-\rangle_\mu |+\rangle_e \\ \psi_4 &= \beta |+\rangle_\mu |-\rangle_e - \alpha |-\rangle_\mu |+\rangle_e \end{aligned} \qquad (8.50)$$

with

$$\begin{aligned}\alpha &= \frac{1}{\sqrt{2}} \sqrt{1 - \frac{x}{\sqrt{1+x^2}}} \\ \beta &= \frac{1}{\sqrt{2}} \sqrt{1 + \frac{x}{\sqrt{1+x^2}}}\end{aligned} \qquad (8.51)$$

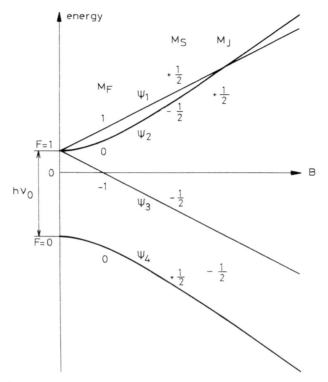

Fig. 8.25 Breit–Rabi diagram for muonium. The energy separations are not to scale.

The Breit–Rabi diagram is shown schematically in Fig. 8.25.

In the Breit–Rabi diagram, only the transition energies can be measured experimentally. In order to clarify this point, which is important for all precession experiments, we will apply the general considerations of Section 3.1.2 to this special case.

8.5.5 Precession of the μ^+ Spin in Muonium

We need to calculate the expectation values of the spin operators σ_x, σ_y, σ_z ($\boldsymbol{\sigma} = 2\mathbf{S}/\hbar$). We note that the σ_i act only on the muon but not the electron wave function. The total wave function at time t can be represented by a superposition of the eigenstates of Eq. (8.50) together with the appropriate time developments

$$\psi(t) = \sum_{i=1}^{4} c_i \psi_i \exp(-i\omega_i t) \tag{8.52}$$

with $\omega_i = E_i/\hbar$ (E_i from Eq. (8.48)). The four constants are determined from the initial value of the polarization (e.g. $\langle\sigma_x\rangle = 1$, $\langle\sigma_y\rangle = \langle\sigma_z\rangle = 0$) and the normalization condition. The polarization in the x-direction (detector position) is then given by

$$P(t) = \langle\psi(t)|\sigma_x|\psi(t)\rangle$$
$$= \sum_{j,k=1}^{4} c_k^* c_j \exp[-i(\omega_j - \omega_k)t]\langle\psi_k|\frac{1}{2}(\sigma_+ + \sigma_-)|\psi_j\rangle \quad (8.53)$$

We will not develop Eq. (8.53) further but instead will discuss some of its implications.

In the time dependence of $P(t)$ only the transition frequencies

$$\omega_{j,k} = \omega_j - \omega_k \quad (8.54)$$

play a role. The selection rules for the possible transitions (note that this is not really a repopulation of energy levels) are obtained by considering the expression $\langle\psi_k|(\sigma_+ + \sigma_-)|\psi_j\rangle$. The terms with $j = k$ in Eq. (8.43) disappear since σ_+ and σ_- have no diagonal elements. In addition one sees from the explicit form of ψ_i in Eq. (8.50) that the transitions $1 \leftrightarrow 3$ and $2 \leftrightarrow 4$ give no contribution to $P(t)$. This can be seen very easily for $1 \leftrightarrow 3$, since the electron wave functions are orthogonal for these states, and σ_+ and σ_- act only on the muon. One can easily see the selection rule '$1 \leftrightarrow 3$ forbidden' since σ is a first-rank tensor (vector) and the states 1 and 3 differ by $\Delta M_F = 2$.

Further treatment of Eq. (8.52) leads to (Celio and Meier, 1983)

$$P(t) = \frac{1}{2}[\cos^2\beta(\cos\omega_{12}t + \cos\omega_{34}t) + \sin^2\beta(\cos\omega_{14}t + \cos\omega_{23}t)] \quad (8.55)$$

with

$$\tan 2\beta = \frac{a}{(-\gamma_e + \gamma_\mu)\hbar B} \quad (8.56)$$

In the Zeeman region where $a \gg (-\gamma_e + \gamma_\mu)\hbar B$, one obtains $\beta = 45°$. Thus in $P(t)$ all four frequencies, each with amplitude 1/4, are present. However, in practice, one usually observes only ω_{12} and ω_{23}, since the other two frequencies are too large to be detected experimentally for normal muonium.

Fig. 8.26 shows the μSR spectrum of muonium in germanium. One clearly sees beats caused by the superposition of two close frequencies ω_{12} and ω_{23}.

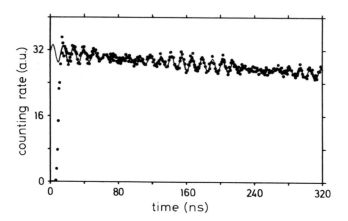

Fig. 8.26 The μSR spectrum of normal muonium in germanium.

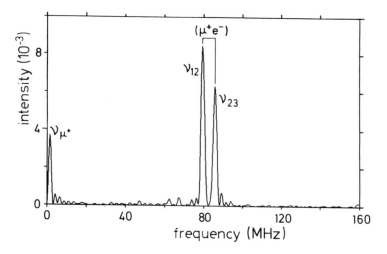

Fig. 8.27 Fourier spectrum of normal muonium in germanium.

Fig. 8.27 shows the corresponding Fourier spectrum. The hyperfine constant a can be determined from the splitting of the lines. For germanium

$$\nu_0 = a/h = 2360 \text{ MHz}$$

This corresponds to 53% of the vacuum value, i.e. in the solid the hyperfine coupling is somewhat reduced.

9
Positron Annihilation

9.1 Principles

The positron (e^+) is the antiparticle of the electron; it is normally produced by β-decay of radioactive nuclei. In a solid, a positron behaves much like a positive muon or proton, and in this respect can be considered a light isotope of hydrogen.

Methods discussed in previous chapters utilize the hyperfine interaction as a tool. This is not the case for the positron annihilation technique. The usefulness of positron annihilation in solid state physics arises because the positron annihilates with an electron and the emitted γ-radiation provides information about the density and velocity distribution of electrons in the solid.

A positron emitted from a β^+ source typically has energy of order 10^5 eV. Such a positron entering the solid loses its energy very rapidly (within ca. 10^{-12} s) until it reaches thermal energies. From this point on, the positron diffuses (like a μ^+ or H^+) until it finally annihilates with an electron through emission of γ-radiation. The dominant emission is two γ-rays with energies of 511 keV. This corresponds to the decay of an electron–positron pair with antiparallel spins. The positron can also meet electrons with parallel spins. This system can, because of the conservation of angular momentum, decay only via three γ-quanta. This decay is less probable by a factor of 372.

Before the positron annihilates it can capture an electron and form a hydrogen-like system (positronium). In the ground state positronium is in either the $^1S_0$ singlet state (para-positronium) or the $^3S_1$ triplet state (ortho-positronium). Without ortho–para conversion, the ratio of these two systems should be 1:3. The singlet state has a mean lifetime of 125 ps, and the triplet state 142 ns. In a solid, the triplet positronium can decay by emission of two γ-rays if it undergoes an ortho to para conversion through a spin-exchange collision with an electron. In metals, no positronium is formed since the charge of the positron is effectively screened by conduction electrons so that no bound state is possible.

Positron Annihilation

In solid state experiments, the following methods are applied:

Angular correlation of the two γ-ray decay

In the center-of-mass system the energy of each annihilation γ-ray is exactly $m_e c^2$ (= 511 keV), and the two γ-rays are emitted in exactly opposite directions. In the laboratory system, this is true only if the e^+e^- pair has no kinetic energy. The energy of thermalized positrons is very small ($E \approx$ 10 meV). However, electrons can have energies up to the Fermi energy ($E \leqslant 10$ eV). This has the consequence that the two γ-rays are not emitted at exactly 180° but at an angle that deviates by θ from 180° (see Fig. 9.1),

$$\theta \approx \frac{p_T}{m_e c} \tag{9.1}$$

Here p_T is the transverse component of the electron–positron pair momentum before annihilation. Thus the deviation of the 180° correlation contains information about the momentum distribution of the electrons.

Energy of the annihilation γ-rays

In addition to a deviation of the γ-ray angular correlation from 180°, the kinetic energy of the e^+e^- pair before annihilation causes a Doppler shift of the two γ-ray energies (see Fig. 9.1). The Doppler shift is $\Delta v/v = v_L/c$ (with the longitudinal center-of-mass velocity $v_L = p_L/2m_e$). For energy $E = m_e c^2$ this leads to a Doppler shift

$$\Delta E = \pm \frac{v_L}{c} E = \pm \frac{c p_L}{2} \tag{9.2}$$

One sees that the energy shift reflects the momentum distribution of the electrons.

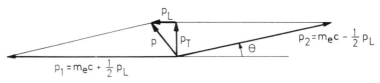

Fig. 9.1 Momentum conservation of the two γ-ray positron annihilation. p is the momentum of the e^+e^- pair, and p_L and p_T the longitudinal and transverse components respectively.

Lifetime of the positron

The decay rate of free positrons in matter depends on the density of electrons at the positron site. For nonrelativistic velocities of the e^+e^- pair, the cross section and consequently the decay rate are proportional to the electron density. The positron can occupy different positions in the solid and therefore is a sensitive probe of electron densities.

9.2 Positron Sources and Experimental Apparatus

9.2.1 Positron Sources

Source requirements for the study of positron annihilation are more easily fulfilled than for Mössbauer and PAC sources. The important properties are a high rate of e^+ radiation and a sufficient positron energy to leave the source and enter the sample. There are a large number of positron sources suitable for these experiments, e.g. ^{22}Na, ^{55}Co, ^{57}Ni, ^{58}Co, ^{64}Cu, ^{68}Ge, and ^{90}Nb. Two commonly used sources, ^{22}Na and ^{64}Cu, are discussed below.

^{22}Na–^{22}Ne

The isotope ^{22}Na has become the standard source for positron annihilation. One reason is the high positron branching ratio (90%), and the other is the conveniently long source lifetime ($t_{1/2} = 2.6$ y). In addition, the ^{22}Na source has another important property: after the β^+-decay having $E_{max} = 544$ keV, a γ-ray having energy $E_\gamma = 1275$ keV is emitted. This γ-ray can be used as a start indicator for lifetime measurements. Several nuclear reactions are suitable for the production of ^{22}Na, e.g. ^{24}Mg(d, α)^{22}Na. Sources of order 10 mCi ($= 3.7 \times 10^8$ decays/s) are usually used. The decay of ^{22}Na is shown in Fig. 9.2.

^{64}Cu–^{64}Ni

The isotope ^{64}Cu can be produced conveniently from ^{63}Cu (natural abundance 68%) by neutron capture through the reaction ^{63}Cu(n, γ)^{64}Cu. However this source is relatively short-lived ($t_{1/2} = 12.7$ h) and has only a small positron branching ratio (19%). The decay of ^{64}Cu is shown in Fig. 9.3.

184   Positron Annihilation

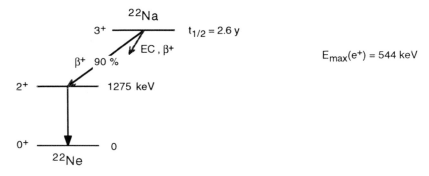

Fig. 9.2 Decay scheme of ^{22}Na (Lederer and Shirley, 1978).

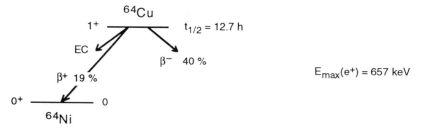

Fig. 9.3 Decay scheme of ^{64}Cu (Lederer and Shirley, 1978).

9.2.2 Experimental Apparatus

Angular correlation experimental apparatus

The set-up of an angular correlation apparatus for two γ-ray annihilation is shown in Fig. 9.4. Positrons leave the source and are stopped in the sample. The annihilation radiation is detected in a fixed and a movable NaI detector for which a small angular range is selected by collimators.

In order to reach an angular resolution of ca. 1 mrad the detectors must be several meters from the sample. The detector signals are amplified and sent to single-channel analyzers to select the 511 keV radiation, and the coincidence count rate is then registered as a function of detector angle (for experimental results, see Fig. 9.8).

Energy spectrum of the annihilation radiation

For the measurements of the γ-ray Doppler shifts (see Eq. (9.2)), γ detectors with very high energy resolution are required, since the energy

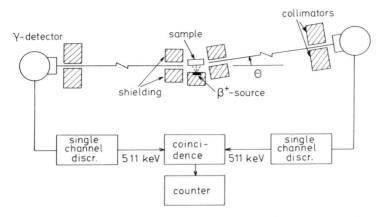

Fig. 9.4 Block diagram of the angular correlation apparatus for two γ-ray positron annihilation.

shift is very small. For electrons with kinetic energy of 10 eV, the Doppler shift that must be detected is smaller than 1.5 keV. In Fig. 9.5, the block diagram of such a set-up is shown.

Lifetime measurements

The set-up for positron lifetime measurements is, in principle, the same as that used for the PAC method. However, the lifetimes that must be measured are in the subnanosecond range whereas for PAC they are in the range of 10 ns to 1 μs. Organic scintillators (plastic materials) which have a very fast time response for γ-radiation are used to achieve the required precision in the time measurements. The disadvantage of these detectors is their poor energy resolution, since the γ-rays are detected only via the Compton effect (see Section 2.4). BaF$_2$ detectors, which have become available in recent years, combine good time resolution with reasonably good energy resolution and are widely used for positron lifetime measurements.

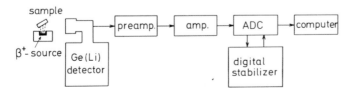

Fig. 9.5 Block diagram of the experimental apparatus used for measurement of positron annihilation radiation Doppler shifts. A stabilized analog-to-digital converter (ADC) is used for the necessarily very precise energy measurements.

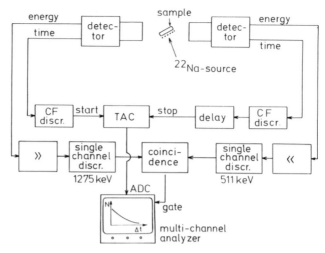

Fig. 9.6 Block diagram of apparatus used for positron lifetime measurements.

The block diagram of the apparatus used for positron lifetime measurements is shown in Fig. 9.6. A ^{22}Na source is used, and the start circuit is tuned to the 1275 keV γ-radiation and the stop circuit to the 511 keV annihilation radiation. Because of the high requirements on time resolution, the time signals are fed directly to the time-to-amplitude converter (TAC) and the energy information is determined afterwards. The TAC signal goes through a linear gate and is rejected if the energy requirements are not fulfilled.

9.3 Angular Correlation of Annihilation Radiation (ACAR) and Fermi Momentum of Conduction Electrons in Metals

As discussed above, the angular correlation of the two γ-ray annihilation radiation provides information about the momentum distribution of the electrons which annihilate with the positrons. We derive below the form of the angular correlation resulting from positron annihilation with free electrons in a metal ($T = 0$ K). In a free electron gas at $T = 0$ K, all states with momentum below the Fermi momentum are occupied, and all higher momentum states are empty. The momentum distribution is a filled sphere with uniform density and radius $\hbar k_F$ ($\hbar k_F$ is the Fermi momentum). A deviation of the angular correlation by an angle θ from 180° is caused by transverse momentum of the conduction electrons (see Eq. (9.1)). The count rate per momentum interval Δp_T of the γ-rays with correlation $180° \pm \theta$ is therefore proportional to the volume of the slice for which $p_T \approx m_e c \theta$ (see Fig. 9.7).

The angular correlation is therefore

$$W(\theta) = \frac{dN}{d\theta} \propto \frac{(\hbar^2 k_F^2 - p_T^2) dp_T}{d\theta} \qquad (9.3)$$

or

$$W(\theta) = \text{const} \times (\hbar^2 k_F^2 - m_e^2 c^2 \theta^2) \qquad (9.4)$$

The angular distribution has a parabolic dependence on θ. The maximum value of $W(\theta)$ occurs at $\theta = 0°$. The coincidence count rate vanishes for angles larger than the maximum angle θ_{max} which is related to the Fermi momentum via

$$\theta_{max} = \frac{\hbar k_F}{m_e c} \qquad (9.5)$$

Fig. 9.8 shows experimental angular correlation data for various metals (Lang et al., 1955). The parabolic dependence is clearly seen; however, the parabolas are superimposed on a broad background. This background originates from positron annihilations with core electrons.

The Fermi momentum in metals can be determined accurately by this method. They are summarized for several metals in Table 9.1 along with other parameters of interest.

In real metals, the Fermi surface is not spherical. The actual Fermi surface is a function of direction in the lattice. It can be measured by angular correlation experiments on single-crystal samples. These experiments yield

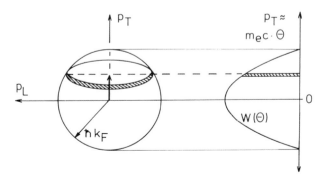

Fig. 9.7 Angular correlation of the annihilation radiation in the free electron gas model.

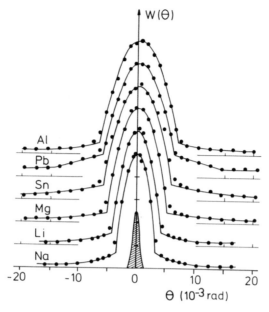

Fig. 9.8 Experimental angular correlations of the two γ-radiation from positron annihilation in various metals. The shaded area represents experimental angular resolution (Lang et al., 1955).

Table 9.1 Density of conduction electrons, Fermi energy E_F, Fermi temperature T_F, and Fermi wave vector k_F for several metals (Frederikse, 1981).

Element	n (10^{22}/cm^3)	E_F (eV)	T_F (10^4 K)	k_F (10^8 cm^{-1})
Li	4.70	4.74	5.51	1.12
Na	2.65	3.24	3.77	0.92
Cu	8.47	7.00	8.16	1.36
Ag	5.86	5.49	6.38	1.20
Au	5.90	5.53	6.42	1.21
Mg	8.61	7.08	8.23	1.36
Ca	4.61	4.69	5.44	1.11
Nb	5.56	5.32	6.18	1.18
Fe	17.0	11.1	13.0	1.71
Cd	9.27	7.47	8.68	1.40
Hg	8.65	7.13	8.29	1.37
Al	18.1	11.7	13.6	1.75
In	11.5	8.63	10.0	1.51
Ti	10.5	8.15	9.46	1.46
Sn	14.8	10.2	11.8	1.64
Pb	13.2	9.47	11.0	1.58

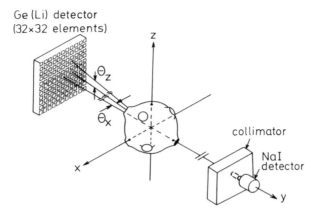

Fig. 9.9 Block diagram of a multidetector arrangement for measurement of Fermi surfaces by positron annihilation. Adapted from Doyama and Hasiguti (1973).

projections of the Fermi surface from which the total Fermi surface can be constructed. The procedure for determining the Fermi surface by angular correlation experiments is illustrated in Fig. 9.9.

9.4 Positron Lifetime and Lattice Defects in Metals

The lifetime of thermalized positrons in a solid depends on the average electron density at the positron site. This fact can be utilized to study lattice defects in the solid. We limit ourselves to metals since positronium formation can be neglected, and the analysis is particularly simple. Lifetime measurements are particularly well-suited for the investigation of vacancies. These are usually negatively charged with respect to the surroundings, since a positive ion has been removed. The positron is attracted to a vacancy and can be captured there. At a vacancy site, the conduction electron density is lower than average, and the positron lifetime is longer than it is when the positron is in a bulk interstitial site.

The concentration of vacancies in a metal can be determined from the ratio of free to captured positrons. In thermal equilibrium, the vacancy concentration c_v is given by

$$c_v = c_0 \exp(-E_v^F/k_B T) \tag{9.6}$$

where c_0 is a prefactor which depends on the entropy, and E_v^F is the energy necessary for formation of a vacancy. E_v^F can be derived by measuring the concentration c_v as a function of temperature. For $E_v^F = 1$ eV, the concentration of vacancies at 1000 K is approximately 10^{-5} per atom. The ratio of the free positron density n_f to that of bound positrons n_v must be determined experimentally. Using model rate equations for capture, release, and decay processes, E_v^F can be derived from this ratio. The rate equations are

$$\frac{dn_f}{dt} = -\lambda_f n_f - \mu_v c_v n_f + \xi_v n_v + N$$
$$\frac{dn_v}{dt} = -\lambda_v n_v + \mu_v c_v n_f - \xi_v n_v \quad (9.7)$$

where N is the injection rate of positrons into the sample, λ_f and λ_v are the annihilation rates of free and vacancy-trapped positrons respectively, and μ_v is the capture rate and ξ_v the detrapping rate from a vacancy (see Fig. 9.10). If positrons are trapped strongly, μ_v is much larger than ξ_v. In the steady state ($dn_f/dt = dn_v/dt = 0$) the fractions of free and bound positrons are

$$f_f = \frac{n_f}{n_f + n_v} = \frac{\lambda_v + \xi_v}{\lambda_v + \xi_v + \mu_v c_v}$$
$$f_v = \frac{n_v}{n_f + n_v} = \frac{\mu_v c_v}{\lambda_v + \xi_v + \mu_v c_v} \quad (9.8)$$

f_f and f_v can each be determined experimentally by determining the two contributions to the annihilation lifetime: $\tau_f = 1/\lambda_f$ and $\tau_v = 1/\lambda_v$. Alternatively the average lifetime $\bar{\tau}$ can be measured, where

$$\frac{1}{\bar{\tau}} = f_f \frac{1}{\tau_f} + f_v \frac{1}{\tau_v} \quad (9.9)$$

With the additional assumption that no detrapping occurs ($\xi_v \approx 0$, valid if binding is strong), the mean lifetime resulting from Eqs. (9.6), (9.8), and (9.9) is

$$\bar{\tau} = \frac{\lambda_v + \mu_v c_0 \exp(-E_v^F/k_B T)}{\lambda_v \lambda_f + \mu_v \lambda_v c_0 \exp(-E_v^F/k_B T)} \quad (9.10)$$

This function is shown in Fig. 9.11. The relevant parameters can be derived from the shape of this curve and the position of the inflection point.

An example of such a measurement for gold (Herlach et al., 1977) is shown in Fig. 9.12. The shape is clearly that discussed above. At low

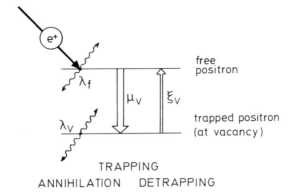

Fig. 9.10 Trapping model for thermalized positrons in a metal. Labels are explained in the text.

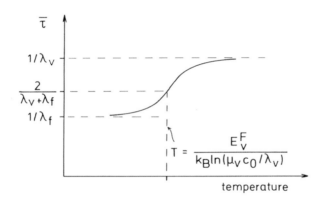

Fig. 9.11 Graphical representation of the mean annihilation lifetime $\bar{\tau}$ (see Eq. (9.10)) as a function of temperature.

temperatures (below 700 K) the vacancy density is negligible, and the positron lifetime reflects annihilation in the bulk crystal. Between approximately 700 K and about 1300 K the equilibrium vacancy concentration increases strongly with temperature, and therefore many positrons are trapped by vacancies where their lifetime is longer than in the bulk. At very high temperature, essentially all positrons annihilate in vacancies, and the average lifetime is expected to become constant. The slight temperature dependence below 700 K can be attributed to lattice expansion and self-trapping effects. At temperatures above 1000 K, multiple vacancy formation may influence the lifetime.

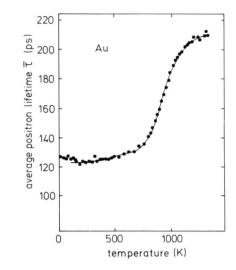

Fig. 9.12 Mean positron lifetime in gold as a function of temperature (Herlach *et al.*, 1977).

The measurement of positron lifetimes can also be used to investigate defects in irradiated materials. An example is shown in Fig. 9.13 for molybdenum which was irradiated at low temperatures with electrons (Eldrup *et al.*, 1975). Following successive isochronal annealing steps, the defects are found to disappear in two stages. This is clearly seen from the resistivity of the sample in Fig. 9.13. In the first stage, the vacancies are assumed to become mobile and agglomerate. In the second stage, the vacancy agglomerates are assumed to dissolve.

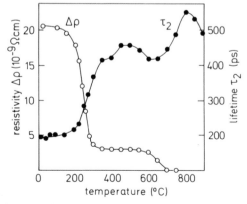

Fig. 9.13 Resistivity change and long positron lifetime τ_2 in electron-irradiated molybdenum as a function of annealing temperature (Eldrup *et al.*, 1975).

Table 9.2 Formation energies E_v^F of single vacancies in several metals (Recknagel et al., 1983).

Metal	Crystal structure	E_v^F (eV) Quenching	E_v^F (eV) Positron annihilation
Al	face-centered cubic	0.66	0.66
Ni		1.60	—
Cu		1.27	1.24
Ag		1.10	1.16
Pt		1.51	—
Au		0.94	0.97
Pb		0.54	0.54
α-Fe	body-centered cubic	> 1.55	1.5
Nb		> 2.7	2.0
Mo		3.2	—
Ta		—	2.2
W		3.9	3.5
Zn	hexagonal	—	0.53
Cd		0.42	0.47

In addition to measuring the resistivity recovery, the long lifetime τ_2 of the e$^+$ annihilation radiation has also been measured (the short-lifetime component is due to free positrons). One sees from Fig. 9.13 that the lifetime strongly increases in the region where the vacancies begin to agglomerate (250 °C). This behavior is attributed to decrease in electron density with increasing vacancy agglomeration. Above 700 °C most of the vacancy clusters dissolve, as can be seen from the decrease in the resistivity. However, the remaining vacancy agglomerates become larger, so the positron lifetime τ_2 again increases above 700 °C. Above 1000 °C all vacancy agglomerates have disappeared.

These two examples involving metallic defects illustrate the importance of positron annihilation for the study of defects. The vacancy formation energies E_v^F obtained by positron annihilation have proven to be among the most reliable. A selection of E_v^F values obtained by positron annihilation and by fast-quenching methods are given in Table 9.2.

10
Neutron Scattering

The neutron, like all particles, has a wavelength given by the de Broglie relation (see Table 10.1). For thermal neutrons ($E_{\text{kin}} \approx 25$ meV) this wavelength is 0.18 nm, and thus is of the same order of magnitude as lattice spacings in solids. Because of this property, neutrons can be used in a manner similar to X-rays for diffraction experiments in solids.

The neutron interacts with condensed matter in two different ways:

(a) Because of the *strong interaction*, the neutron is scattered by nuclei in the sample. Thus they may be used in a manner similar to X-rays for determining the spatial arrangement of atoms in the solid.
(b) Because of its *magnetic moment*, the neutron can also be scattered by magnetic atoms in the solid, so a determination of magnetic structure in solids is also possible using neutrons.

Neutrons are particularly useful for investigating dynamic processes (diffusion, vibrations, etc.) in condensed matter. This is true because the energy of neutrons is relatively low compared to X-rays of the same wavelength, so only modest energy resolution is required to separate elastic from inelastic and quasi-elastic neutron scattering. An illustration is given in the following example.

Table 10.1 Properties of the neutron

Mass	939.573 MeV/c^2
Spin	1/2
Decay	β-decay: $n \to p + e^- + \bar{\nu}_e$
Lifetime	888.60(35) s
g-factor	$g_S = -3.82638$
Wavelength	$\lambda = \dfrac{h}{p} = \dfrac{h}{\sqrt{2m_n E}}$; $\lambda(\text{nm}) = \dfrac{0.0286}{\sqrt{E(\text{eV})}}$
	$\lambda = 0.18$ nm for $E = 25$ meV (≈ 290 K)
Interaction with matter	(a) nuclear interaction ($\to$ nuclear force)
	(b) magnetic interaction ($\to$ g-factor)

For X-rays, the wavelength of 0.1 nm, a typical value in diffraction measurements, corresponds to an energy of 12 keV ($\lambda(\text{nm}) = 1.24/E(\text{keV})$). For neutrons, this wavelength corresponds to an energy of 80 meV ($\lambda(\text{nm}) = 0.0286/\sqrt{E(\text{eV})}$). Elementary excitations in the solid, such as phonons or magnons, have energy of order 0.1 eV or smaller. In X-ray diffraction, inelastic measurements with this precision would be extremely difficult (requiring precision better than 0.1 eV in 12 keV), whereas for neutrons, inelastic energy transfers are of the same order of magnitude as the neutron kinetic energy.

The properties of neutrons, production of monochromatic neutron beams, detection of neutrons, measurement of neutron energies, and neutron diffraction with applications and examples are discussed below. A detailed description of neutron scattering can be found in Bacon (1975).

10.1 Properties of Neutrons and Production of Neutron Beams

The neutron decays by β-emission into a proton, an electron, and an antineutrino

$$n \rightarrow p + e^- + \bar{\nu}_e$$

The β-decay is parity-violating; the antineutrino $\bar{\nu}_e$ has right-helicity, the electron is left-circularly polarized.

For production of neutrons, either nuclear reactors or particle accelerators are used. The latter is called a spallation neutron source, since the neutrons are created by spallation (splitting up) of nuclei.

Nuclear reactors

The typical reaction in the reactor is

$$^{235}\text{U} + n_{\text{therm}} \rightarrow A + B + 2.3\,n$$

where A and B are two fission fragments of the original ^{235}U nucleus, and n_{therm} is a neutron with thermal energy. The neutrons from the nuclear fission are slowed by a moderator (e.g. H_2O) to thermal energy. They can then initiate further nuclear fission (chain reaction). A few neutrons are allowed to escape from the reactor core through openings and are then available for experiments. Typical powers of research reactors are 10 to 100 MW (the high-flux reactor in Grenoble runs at 57 MW, the medium-flux reactors in Jülich and Berlin at 10 MW).

The energy distribution of the neutrons which leave the nuclear reactor depends on the temperature of the moderator. The neutron energy distributions available at beam lines in Grenoble are shown in Fig. 10.1. The 'cold' neutrons are produced by passing neutrons from the core through a liquid deuterium (at temperature $T = 25$ K) moderator and then guiding them to the experimental set-up. To produce 'hot' neutrons, one may use a graphite moderator at a temperature of 2000 K.

Spallation neutron sources

Here the neutrons are produced by bombarding heavy elements with high-energy protons ($E_p \approx 800$ MeV). A typical nuclear reaction is

$$p + {}^{238}U \rightarrow \text{spallation product} + x\,n$$

The number of neutrons per proton depends strongly on the proton energy and the target material. For 800 MeV protons on uranium, one obtains an average of 28 neutrons per proton. If one uses tantalum instead of uranium to avoid breeding plutonium, the neutron yield drops by one half.

After production, the neutrons are moderated, as in the nuclear reactors, before they are used for scattering experiments. Pulsed accelerators are

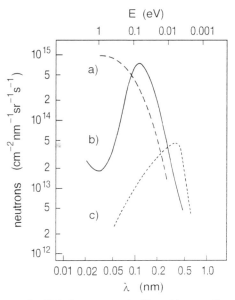

Fig. 10.1 Neutron flux at the high-flux reactor in Grenoble as a function of wavelength for different moderator temperatures: (a) 2000 K, (b) 300 K, (c) 25 K (Bée, 1988).

often used for the spallation, and pulsed neutron beams are produced. Pulsed beams have a particular advantage for time-of-flight measurements, since the mechanical choppers (see below) are not necessary.

Monochromators

Mechanical monochromators consist of a solid cylinder of a material with strong neutron absorptivity that has a hole through which the neutrons are allowed to pass (see Fig. 10.2). Neutrons with the correct velocity can pass through the curved hole of the rotating cylinder. All others hit the cylinder and are absorbed. The monochromatic neutron beam is emitted in pulses by this device, a desirable feature for time-of-flight experiments.

An alternative method for producing monochromatic neutron pulses is to have several slitted disks that rotate on a common axis. By choosing an appropriate rotation velocity and slit offset in the different disks, neutrons of a given energy can be selected. These devices are called choppers.

The *crystal monochromator* makes use of the Bragg condition for

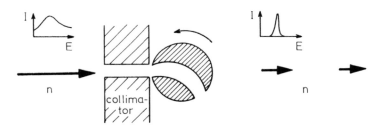

Fig. 10.2 Principle of the mechanical monochromator. Only neutrons of a certain velocity (energy) can pass through the hole on a straight path. All others are absorbed. The emitted beam is pulsed.

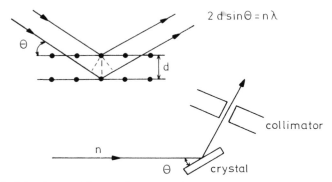

Fig. 10.3 Principle of a crystal monochromator. A neutron beam impinges on the crystal at angle θ and is diffracted by the crystal planes. For $2d \sin \theta = n\lambda$, Bragg reflection occurs.

diffracted particles (see Fig. 10.3). The monochromator crystal should have a small neutron absorption coefficient. Pyrolytic graphite, copper, germanium, silicon, and berylium are commonly used.

10.2 Detection of Neutrons

Since a neutron has no charge, it cannot be detected by ionization processes. Therefore, neutrons are detected through nuclear reactions producing charged products with large kinetic energies (exothermic reactions). The resulting charged particles can then be detected. Common reactions are

$$n + {}^{10}B \xrightarrow{93\%} {}^{7}Li^* + {}^{4}He \rightarrow {}^{7}Li + {}^{4}He + \gamma \quad E_\gamma = 0.48\,\text{MeV}$$
$$E_{kin}({}^{7}Li + {}^{4}He) = 2.3\,\text{MeV}$$
$$\xrightarrow{7\%} {}^{7}Li + {}^{4}He \quad E_{kin}({}^{7}Li + {}^{4}He) = 2.78\,\text{MeV}$$

and

$$n + {}^{3}He \rightarrow p + {}^{3}H \quad E_{kin}(p + {}^{3}H) = 0.77\,\text{MeV}$$

Boron trifluoride (BF$_3$) *detector*

The detector consists of a proportional counter filled with BF$_3$ gas enriched with the ^{10}B isotope (natural abundances 80% ^{11}B, 20% ^{10}B). The gas in the counter tube has a pressure above 1 bar in order to provide enough ^{10}B nuclei for neutron detection. The ^{7}Li and ^{4}He nuclei emitted in the nuclear reaction ionize the gas. The charges created by the ionization are separated by the applied high voltage and detected as a voltage pulse at the load resistor. The detection efficiency for thermal neutrons with energies below 1 eV is approximately 50%. For neutron energies above 1 eV, the efficiency decreases drastically. Fig. 10.4 shows a typical pulse height spectrum.

^{3}He detector

In this detector, the counting gas consists of ^{3}He. Since the thermal neutron capture cross section of ^{3}He is 40% larger than for ^{10}B ($\sigma(^{10}B) = 3840\,\text{b}$, $\sigma(^{3}He) = 5500\,\text{b}$), these detectors can be smaller. The smaller dimensions have the advantage of shorter collection time for electrons produced by ionization, so the time resolution is better.

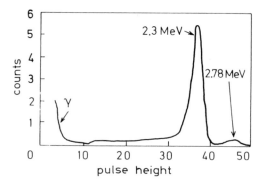

Fig. 10.4 Pulse height spectrum of a BF_3 counter irradiated by thermal neutrons (Beckurts and Wirtz, 1964). The lines at 2.3 MeV and 2.78 MeV correspond to the total kinetic energy of ^7Li and ^4He.

The arrival time at the detector but not the energy of the neutrons can be determined using these detectors. For energy measurement one uses other techniques, e.g. time-of-flight measurement. There the neutron energy is determined from the flight time of the neutron between sample and detector. The experimental set-up is shown in Fig. 10.5.

A crystal spectrometer can also be used to measure neutron energies. The energy of scattered neutrons can be determined by Bragg diffraction (see Fig. 10.6) from a crystal of known lattice spacing (analyzer).

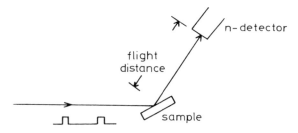

Fig. 10.5 Block diagram of a time-of-flight spectrometer.

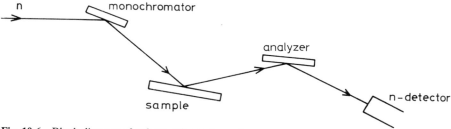

Fig. 10.6 Block diagram of a three-axis spectrometer.

10.3 Theory of Neutron Scattering

10.3.1 Neutron Scattering by Nuclei

We consider elastic scattering of neutrons on bound nuclei in the solid. In this case, as in the Mössbauer effect, the recoil can be neglected, so that the energy of the scattered neutrons is equal to the energy of the incident neutrons. Thus

$$|k| = |k'| = \frac{2\pi}{\lambda} \tag{10.1}$$

where k and k' are the wave vectors of the incident and outgoing neutrons. The total wave function of neutrons at large distances from the scattering center is

$$\psi(r, \theta, \phi) = A\left[\exp(ikz) + f(\theta, \phi)\frac{\exp(ikr)}{r}\right] \tag{10.2}$$

where $A\exp(ikz)$ is the incident plane wave, and $A[f(\theta, \phi)\exp(ikr)]/r$ is the outgoing spherical wave; A is a normalization constant, and $f(\theta, \phi)$ the scattering amplitude. For thermal neutrons, the wavelength is large compared to the range of the nuclear potential so that only s-wave scattering contributes. In this case $f(\theta, \phi)$ is a constant. Neglecting absorption, one has

$$f(\theta, \phi) = -b \tag{10.3}$$

where b is a positive or negative real number; b is called the scattering length.

The differential cross section $d\sigma/d\Omega$ is defined to be the ratio of the number of particles scattered into the solid angle element $d\Omega$ to the incident particle flux density. Using Eqs. (10.2) and (10.3), one obtains

$$\frac{d\sigma}{d\Omega} = |f(\theta, \phi)|^2 = b^2 \tag{10.4}$$

or for the total cross section

$$\sigma = \int \frac{d\sigma}{d\Omega} d\Omega = 4\pi b^2 \tag{10.5}$$

In Eqs. (10.4) and (10.5) one sees that the scattering length b describes the

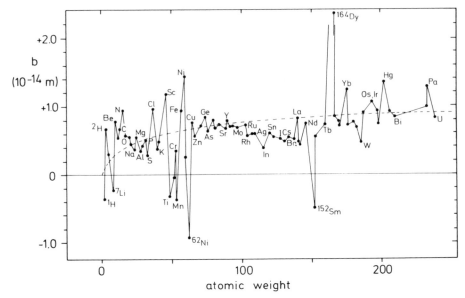

Fig. 10.7 Scattering length as a function of atomic mass. The dashed line corresponds to potential scattering (Jeffrey, 1981).

strength of the scattering; b depends on the radius and depth of the nuclear potential and, in general, is different for each isotope. In Fig. 10.7 several scattering length values are given. If no isotope is indicated, b is the mean value of all isotopes of an element. The general trend of b as a function of atomic weight is indicated by the dashed line in Fig. 10.7. One sees that, in contrast to X-ray scattering where the cross section increases linearly with nuclear charge Z, there is no strong mass dependence for neutron scattering. Thus, both light and heavy elements can be detected by neutron scattering with almost equal sensitivity. The scattering length is an individual property of each isotope.

10.3.2 *Neutron Diffraction by Condensed Matter*

When considering scattering of slow neutrons by condensed matter, all contributions of the individual scattering centers must be summed coherently. This means that differences in the path lengths must be taken into account in the phase factor of the plane waves (see Fig. 10.8).

The phase difference between neutrons scattered at point r_n and neutrons scattered at the origin is

$$\boldsymbol{k} \cdot \boldsymbol{r}_n - \boldsymbol{k}' \cdot \boldsymbol{r}_n = (\boldsymbol{k} - \boldsymbol{k}') \cdot \boldsymbol{r}_n = -\boldsymbol{q} \cdot \boldsymbol{r}_n \qquad (10.6)$$

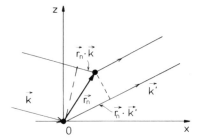

Fig. 10.8 Geometrical construction illustrating the path length differences between neutrons scattered at the point r_n and neutrons scattered at the origin. From this path difference, the phase shift is easily calculated.

where

$$k - k' = \Delta k := -q \tag{10.7}$$

The relationship between the scattering vector q and the scattering angle 2θ can be read off Fig. 10.9. One has

$$\sin\theta = \frac{|q|}{2|k|} \tag{10.8}$$

The coherent summation of all scattering contributions in direction k' gives for the scattering amplitude f_c (c indicates condensed matter)

$$f_c = \sum_n b_n \exp(-i q \cdot r_n) \tag{10.9}$$

where r_n indicates the position of the scattering center n.

For the moment, we make the following assumptions (we abandon these assumptions later and discuss the consequences on the scattering cross sections):

(a) All atoms are of the same nuclear isotope, and the nuclear spin is zero.
(b) The unit cell contains only one atom.
(c) The nuclei are motionless.

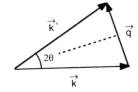

Fig. 10.9 Relationship between the scattering vector q and the scattering angle 2θ for elastic scattering ($|k'| = |k|$).

Because of assumptions (a) and (b), all atoms are equivalent. From this it follows that their scattering lengths b_n are equal ($b_n = b$), and we obtain

$$\frac{d\sigma}{d\Omega} = |f_c|^2 = b^2 \left| \sum_n \exp(-i\boldsymbol{q} \cdot \boldsymbol{r}_n) \right|^2 \tag{10.10}$$

If the origin is located at a lattice point,

$$\boldsymbol{r}_n = p_n \boldsymbol{a} + q_n \boldsymbol{b} + r_n \boldsymbol{c} \tag{10.11}$$

where p_n, q_n, r_n are integers, and $\boldsymbol{a}$, $\boldsymbol{b}$, $\boldsymbol{c}$ are unit-cell vectors. The reciprocal lattice vectors have the following form,

$$\boldsymbol{G}_{hkl} = h\boldsymbol{A} + k\boldsymbol{B} + l\boldsymbol{C} \tag{10.12}$$

with integer h, k, l (Miller indices) and

$$\boldsymbol{A} = \frac{2\pi(\boldsymbol{b} \times \boldsymbol{c})}{\boldsymbol{a} \cdot (\boldsymbol{b} \times \boldsymbol{c})} \qquad \boldsymbol{B} = \frac{2\pi(\boldsymbol{c} \times \boldsymbol{a})}{\boldsymbol{a} \cdot (\boldsymbol{b} \times \boldsymbol{c})} \qquad \boldsymbol{C} = \frac{2\pi(\boldsymbol{a} \times \boldsymbol{b})}{\boldsymbol{a} \cdot (\boldsymbol{b} \times \boldsymbol{c})} \tag{10.13}$$

In Eqs. (10.11)–(10.13) $\boldsymbol{G}_{hkl} \cdot \boldsymbol{r}_n$ is always an integer multiple of 2π. Using this fact, it follows that

$$\left| \sum_n \exp(-i\boldsymbol{G}_{hkl} \cdot \boldsymbol{r}_n) \right|^2 = \left| \sum_n \exp(-i2\pi) \right|^2 = N^2 \tag{10.14}$$

with N the number of atoms. If, in addition,

$$\boldsymbol{q} = \boldsymbol{G}_{hkl} \tag{10.15}$$

there is constructive interference (Bragg condition). For $\boldsymbol{q}$ values which are far away from a reciprocal lattice vector, the contributions of the different atoms in Eq. (10.10) sum to zero. For $\boldsymbol{q}$ values near $\boldsymbol{G}_{hkl}$ one obtains nonzero contributions to the cross section. The width of the Bragg line is determined by these processes. The width is proportional to $1/N$. With these considerations, the differential cross section is found to be

$$\frac{d\sigma}{d\Omega} \propto b^2 N \delta(\boldsymbol{q} - \boldsymbol{G}_{hkl}) \tag{10.16}$$

where δ is the Kronecker delta symbol. To find the absolute size of the cross section, we would need to determine how many reflections fall into the solid angle $d\Omega$.

We now drop the first assumption (all isotopes equal and no nuclear spin). For nuclei with nonzero spin, scattering can occur for different relative orientations of the neutron and nuclear spins; these processes have different scattering lengths. As emphasized earlier, the same is true for scattering by different isotopes. Thus, in general, the scattering potential is not completely periodic although the atomic arrangement is periodic. For statistically distributed spin directions and isotopes, we have a statistically fluctuating potential. The potential fluctuations around the average do not contribute to the coherent scattering and lead to an isotropic (angular-independent) scattering. This part is called incoherent neutron scattering.

Coherent neutron scattering arises because the average periodic potential is nonzero even when nuclei are isotopically mixed and have different spin directions. This leads to interference (Bragg diffraction). The coherent neutron scattering amplitude is proportional to the average scattering length

$$\bar{b} = \frac{1}{N}\sum_{n=1}^{N} b_n \tag{10.17}$$

One obtains

$$\left(\frac{d\sigma}{d\Omega}\right)_{coh} \propto \bar{b}^2 N \delta(\boldsymbol{q} - \boldsymbol{G}_{hkl}) \tag{10.18}$$

For incoherent scattering deviations from the average are important,

$$\overline{b^2} - \bar{b}^2 = \frac{1}{N}\left[\sum_{n=1}^{N} b_n^2 - \left(\sum_{n=1}^{N} b_n\right)^2\right] \tag{10.19}$$

One obtains

$$\left(\frac{d\sigma}{d\Omega}\right)_{inc} = N(\overline{b^2} - \bar{b}^2) \tag{10.20}$$

The incoherent scattering cross section is angular-independent. In Table 10.2 a few values for coherent and incoherent scattering cross sections are given. One sees that hydrogen scatters mainly incoherently whereas the coherent cross section dominates for deuterium. For this reason, deuterium is better suited for structure investigations.

Table 10.2 Scattering lengths of a few atoms. If no isotopic mass number is indicated, the natural abundance is assumed (Bacon, 1975).

Nucleus	Nuclear spin	σ_{coh} (barn)	σ_{inc} (barn)
^{1}H	1/2	1.8	79.7
^{2}H	1	5.6	2.0
^{9}Be	3/2	7.5	0.0
^{27}Al	5/2	1.5	0.0
Ca		2.8	0.4
Fe		11.3	0.5
Ni		13.3	4.7
Cu		7.3	1.2
Zn		4.1	0.1
Pd		4.5	0.3
W		2.9	2.8
Pb		11.1	0.3

Debye–Waller factor

If the condition that all atoms are motionless is abandoned, we obtain a Debye–Waller factor which gives the reduction of the reflection intensities by lattice vibrations. This factor was derived in Chapter 4 (Mössbauer effect). It can be directly introduced here by replacing k in Eq. (4.13) by q. One obtains, with the simplifying assumption made in Chapter 4,

$$f_{DWF} = |\exp(-W)|^2 = \exp(-q^2 \langle u^2 \rangle / 3) \qquad (10.21)$$

where q is the magnitude of the scattering vector, and $\langle u^2 \rangle$ the average square displacement of the atoms. The intensity of a Bragg reflection is then

$$I_{hkl} = I^0_{hkl} \exp(-2W) \qquad (10.22)$$

Geometrical structure factor

If we also drop the assumption that the unit cell contains only one atom, we have to include scattering by atoms in the unit cell when calculating the strength of a Bragg reflection. Designating the vectors p_i as the position of the unit cell and ρ_i as the position of atoms within the unit cell, one can rewrite Eq. (10.9) in the following way (including the Debye–Waller factor)

$$f_c = \sum_n \exp(-i\mathbf{q} \cdot \mathbf{p}_n)\left[\sum_j \bar{b}_j \exp(-W_j)\exp(-i\mathbf{q} \cdot \boldsymbol{\rho}_j)\right] \quad (10.23)$$

where the sum over n is over all unit cells and the sum over j is over all atoms in the nth unit cell. The Debye–Waller factor is generally different for different atoms in the unit cell. Restricting $\mathbf{q}$ to be an allowed Bragg reflection $\mathbf{G}_{hkl}$, one obtains

$$F_{hkl} = \sum_j \bar{b}_j \exp(-W_j)\exp(-i\mathbf{G}_{hkl} \cdot \boldsymbol{\rho}_j) \quad (10.24)$$

With

$$\begin{aligned} \mathbf{G}_{hkl} &= h\mathbf{A} + k\mathbf{B} + l\mathbf{C} \\ \boldsymbol{\rho}_j &= u_j\mathbf{a} + v_j\mathbf{b} + w_j\mathbf{c} \end{aligned} \quad (10.25)$$

where u_j, v_j, and w_j are real numbers between 0 and 1, and

$$\begin{aligned} \mathbf{a} \cdot \mathbf{A} &= \mathbf{b} \cdot \mathbf{B} = \mathbf{c} \cdot \mathbf{C} = 2\pi \\ \mathbf{a} \cdot \mathbf{B} &= \mathbf{a} \cdot \mathbf{C} = \ldots = 0 \end{aligned} \quad (10.26)$$

one obtains

$$F_{hkl} = \sum_j \bar{b}_j \exp(-W_j)\exp[-2\pi i(u_j h + v_j k + w_j l)] \quad (10.27)$$

F_{hkl} is called the geometrical structure factor. The intensity of the Bragg reflection is then

$$I_{hkl} \propto |F_{hkl}|^2 \quad (10.28)$$

10.4 Elastic Neutron Scattering

Analogous to X-ray scattering, elastic neutron scattering is used for elucidation of crystal structure. As mentioned earlier, the advantage of neutron diffraction is that light elements (e.g. H) can be detected and that neutron scattering is sensitive to the magnetic structure. An example of each topic is given below.

Neutron Scattering

EXAMPLE Lattice structure of NaH

The NaH crystal consists of two face-centered cubic sublattices illustrated in Fig. 10.10. This lattice structure is, of course, determined by the diffraction experiment, but we will take the simpler approach of simply demonstrating that this structure agrees with observed reflections.

The NaH unit cell contains four Na and four H atoms at the following positions

Na			H		
u	v	w	u	v	w
0	0	0	1/2	1/2	1/2
0	1/2	1/2	1/2	0	0
1/2	0	1/2	0	1/2	0
1/2	1/2	0	0	0	1/2

The structure factor (Eq. (10.27)) without the Debye–Waller factor has the following form:

$$F_{hkl} = b(\text{Na})\{1 + \exp[-i\pi(k+l)] + \exp[-i\pi(h+k)] \\ + \exp[-i\pi(h+l)]\} \\ + b(\text{H})\{\exp[-i\pi(h+k+l)] + \exp[-i\pi h] + \exp[-i\pi k] \\ + \exp[-i\pi l]\} \quad (10.29)$$

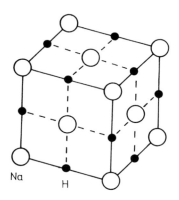

Fig. 10.10 Model of the NaH crystal.

The F_{hkl} are

h, k, l even: $\quad F = 4b(Na) + 4b(H)$

h, k, l odd: $\quad F = 4b(Na) - 4b(H)$

h even; k, l odd: $\quad F = 0$

h, k even; l odd: $\quad F = 0$

For X-ray diffraction, $b(H)$ is negligible compared to $b(Na)$, and one obtains reflection if either all hkl are even or all are odd. For neutron scattering, $b(Na) \approx -b(H)$ (see Fig. 10.7), so that no Bragg reflections are expected for hkl even. The observed reflections (see Fig. 10.11) correspond to these predictions, and thus confirm the proposed structure.

EXAMPLE Crystal structure of the high-temperature superconductor $YBa_2Cu_3O_7$

$YBa_2Cu_3O_7$ is a typical example of the newly discovered copper oxide superconductors; its transition temperature to superconductivity is 93 K, i.e. above the temperature of liquid nitrogen (77 K). Thus the cooling problems are greatly reduced compared to the liquid helium cooling required for classical superconductors.

The crystal structure of $YBa_2Cu_3O_7$ was first determined by neutron scattering. This is not surprising, since the light element oxygen has to

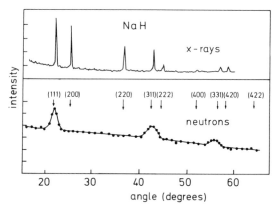

Fig. 10.11 X-ray and neutron scattering by face-centered cubic NaH. Adapted from Shull *et al.* (1948) and Hellwege (1976).

be detected in the presence of heavy elements such as Ba. This is difficult with X-rays.

Fig. 10.12 shows the neutron diffraction pattern of the $YBa_2Cu_3O_7$ powder sample. The observed Bragg reflections are indexed according to an orthorhombic unit cell. The analysis gives for the lattice constants: $a = 0.382$ nm, $b = 0.388$ nm, and $c = 1.175$ nm. The fact that c is approximately three times larger than b and that a is somewhat smaller than b is reflected in the triplet indexed (003), (010), (100) at a scattering angle of approximately 38°.

The X-ray diffraction pattern of $YBa_2Cu_3O_7$ has a shape qualitatively similar to the neutron pattern shown in Fig. 10.12; in particular the same reflections are found, but the line intensities, which are determined by the structure factor (Eq. (10.24)) are different. In X-ray scattering, the contribution from oxygen is very weak, whereas for neutron scattering, it is actually somewhat larger than for barium (compare Fig. 10.7). The crystal structure shown in Fig. 10.13 was derived from a detailed fit of the calculated line intensities to the measured neutron diffraction data. Important structural properties include the CuO_2 planes above and below the Y atoms and the CuO chains along the b axis of the unit cell.

The electrical conductivity and the superconductivity are confined largely to the CuO_2 planes whereas the conductivity perpendicular to these planes is much smaller. This anisotropy is one of the problems in technical applications of these materials, since optimum transport

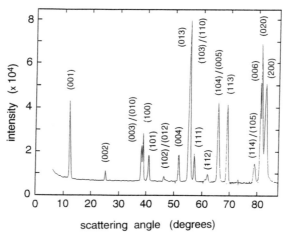

Fig. 10.12 Neutron diffraction pattern of a $YBa_2Cu_3O_7$ powder sample. The indexing (hkl) refers to an orthorhombic unit cell.

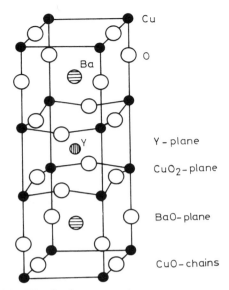

Fig. 10.13 Unit cell of the $YBa_2Cu_3O_7$ superconductor.

behavior requires either single crystals or at least c-axis-aligned materials.

The oxygen in the CuO chains can be removed easily by annealing in vacuum. With appropriate choice of oxygen partial pressure and temperature, any arbitrary oxygen concentration between O_6 (all O atoms from the chain are removed) and O_7 (all O-atom sites in the chains are occupied) can be obtained. The fact that the O atoms are removed from the chains and not from other positions was discovered by neutron diffraction. The result is obtained by precise comparison of the measured line intensities with the calculated ones where the O concentration at each lattice position is varied.

EXAMPLE Magnetic structure of $YBa_2Cu_3O_6$

The removal of oxygen from the CuO chains continuously reduces the superconducting transition temperature; at $O_{6.4}$, superconductivity disappears completely. At approximately the same oxygen stoichiometry, an onset of magnetic ordering in this material is observed. The magnetism is fully developed at $O_{6.0}$.

The fact that $YBa_2Cu_3O_6$ shows magnetic ordering at a reduced oxygen content was first discovered in μSR investigations by the appearance of μSR precessions in zero external field. Although these

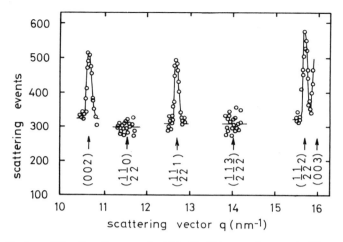

Fig. 10.14 Neutron diffraction diagram of a $YBa_2Cu_3O_{6.15}$ powder sample (Tranquada et al., 1988) in the scattering vector range between the (002) and (003) reflections. The observation of the half-integer reflection $(\frac{1}{2}\frac{1}{2}1)$ is a clear indication of a superstructure which in the present case is of magnetic origin. The non-observation of the $(\frac{1}{2}\frac{1}{2}0)$ and $(\frac{1}{2}\frac{1}{2}\frac{3}{2})$ reflections gives information on the arrangement of the magnetic moment in the unit cell (see text).

experiments showed unambiguously the presence of magnetic ordering, it was not possible with μSR to determine the magnetic structure. The magnetic structure was eventually determined by neutron diffraction.

Fig. 10.14 shows the neutron diffraction pattern for a $YBa_2Cu_3O_{6.15}$ powder sample in the angular range between the (002) and (003) reflections. There is a clear additional reflection which requires half-integer indexing. This line was attributed to magnetic scattering (later confirmed with polarized neutrons). Analysis of the total spectrum, in particular of the magnetic reflections, demonstrates that the Cu atoms in the CuO_2 planes of this substance are antiferromagnetically ordered and have magnetic moments of approximately 0.6 μ_B.

The interplay of magnetism and superconductivity in these materials and the fact that slight changes in oxygen stoichiometry convert superconducting materials to magnetic materials and vice versa are among the most interesting subjects in high-temperature superconductivity research. Whether the tendency to magnetic order is associated with the superconducting mechanism is presently an open question.

10.5 Quasi-Elastic Neutron Scattering

A neutron which is scattered by a rapidly diffusing particle will absorb or give up some energy during the scattering process. This is actually an inelastic process since the neutron energy is changed. However, since the energy transfer is small and leads only to a broadening of the elastic line, it is called quasi-elastic scattering. For a detailed description of quasi-elastic scattering, see Lechner (1983).

The space-time behavior of a diffusing particle in this context is best described by the autocorrelation function $G_s(r, t)$ which gives the semiclassical probability that a particle at time t is at position r if that particle was at the origin at $t = 0$. Van Hove (1954) has shown that the autocorrelation function is associated with the cross section for incoherent neutron scattering $d^2\sigma_{inc}/d\Omega\, d\omega$ by the following expression

$$\frac{d^2\sigma_{inc}}{d\Omega\, d\omega} = \frac{\sigma_{inc}}{4\pi} \frac{k'}{k} S_{inc}(q, \omega) \qquad (10.30)$$

with

$$S_{inc}(q, \omega) = \iint G_s(r, t) \exp[i(q \cdot r - \omega t)]\, d^3r\, dt \qquad (10.31)$$

where $\hbar\omega$ is the transferred energy, $\hbar q$ the transferred momentum, σ_{inc} the total incoherent cross section (see Table 10.2), and $\hbar k$ and $\hbar k'$ the neutron momenta before and after scattering. S_{inc} is called the incoherent scattering function. According to Eq. (10.31) it is just the Fourier transform of the autocorrelation function $G_s(r, t)$. Eq. (10.30) implies that the incoherent scattering cross section per unit solid angle and energy is a measure of the incoherent scattering function S_{inc}. Eqs. (10.30) and (10.31) arise from the dependence of the scattering cross section on the spectral distributions of the momenta and frequencies of the autocorrelation function $G_s(r, t)$. Expressions similar to Eqs. (10.30) and (10.31) exist also for the coherent scattering cross section (intensities in the Bragg reflection). In this case it is necessary to introduce the total correlation function $G(r, t)$ in which no distinction between a given atom and another equivalent atom is made. For measurement of diffusion by quasi-elastic neutron scattering, we need only the incoherent scattering (intensity between the Bragg reflections).

It is sometimes useful to introduce an intermediate function $I_s(q, t)$ in which $G_s(r, t)$ is Fourier transformed with respect to space but not with respect to time

$$I_s(q, t) = \int G_s(r, t) \exp(iq \cdot r)\, d^3r \qquad (10.32)$$

EXAMPLE Diffusion in water

Quasi-elastic neutron scattering is often used to study the dynamics of fluids or motion in solids. A classical example is the diffusion of water molecules in water.

We choose the coordinate system so that a water molecule is at the origin at $t = 0$ and examine the probability that this particle is at r at time t later. This probability corresponds to the autocorrelation function $G_s(r, t)$. In classical (continuous) diffusion, Fick's second law

$$\frac{\partial G_s(r, t)}{\partial t} = D \Delta G_s(r, t) \tag{10.33}$$

can be applied and must be solved with the boundary conditions $G_s(r, 0) = \delta(r)$. D is the self-diffusion coefficient, and Δ the Laplace operator. The normalized ($\int G_s(r, t) \, d^3 r = 1$) solution is

$$G_s(r, t) = (4\pi Dt)^{-3/2} \exp\left(-\frac{r^2}{4Dt}\right) \tag{10.34}$$

This equation implies that the probability distribution of the particle at time t is a bell-shaped curve centered at the original site of the particle. The width of the curve is determined by Dt. In Fig. 10.15 the calculated bell-shaped curves are displayed for the realistic case of

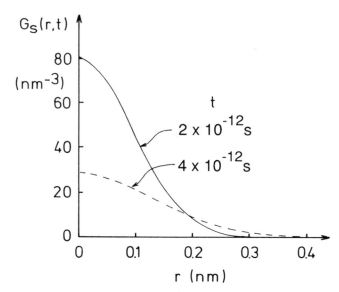

Fig. 10.15 Radial dependence of the autocorrelation function calculated using Eq. (10.34) for $D = 2.13 \times 10^{-5}$ cm^2 s^{-1} (realistic value for water at 25 °C).

water self-diffusion. From these data one can see that a water molecule diffuses a distance of order 0.1 nm in a time of order 10^{-12} s (the values for D were obtained from neutron scattering as shown below). For the very small distances considered here, it seems doubtful that the assumption of continuous diffusion is a good model, since discontinuous jumps are likely to be important for short times.

The scattering function $S_{inc}(q, \omega)$, which can be determined in neutron scattering, is obtained from a Fourier transformation of Eq. (10.34),

$$S_{inc}(q, \omega) = \int\int (4\pi Dt)^{-3/2} \exp\left(-\frac{r^2}{4Dt}\right) \exp\left[i(q \cdot r - \omega t)\right] d^3r \, dt$$

$$= \frac{1}{\pi} \frac{Dq^2}{\omega^2 + (Dq^2)^2} \quad (10.35)$$

For fixed q this corresponds to a Lorentz distribution with full width at half-maximum (FWHM)

$$\Delta\omega = 2Dq^2 \quad (10.36)$$

Fig. 10.16 shows the measured scattering function for water at 25 °C for different magnitudes of the scattering vector q. The diffusion coefficients can be determined from the widths of these curves using Eq. (10.36). After correcting for experimental resolution, one obtains

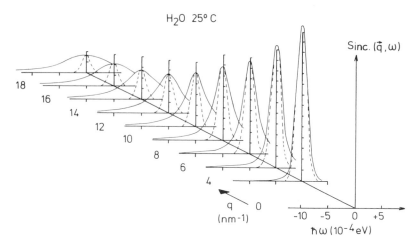

Fig. 10.16 Measured scattering function $S_{inc}(q, \omega)$ for water at 25 °C for different values of q. The dashed curves give the experimental resolution (Sakamoto et al., 1962).

an energy width of $\Delta E = \hbar\Delta\omega = 2.8 \times 10^{-4}$ eV at $q = 10$ nm^{-1}. Using Eq. (10.36), this corresponds to $D = 2.13 \times 10^{-5}$ cm^2 s^{-1} for water at 25 °C.

In Fig. 10.16 one can see that the width of the quasi-elastic line increases with q. According to Eq. (10.36) $\Delta\omega$ should be proportional to q^2. However, one finds experimentally that there are clear deviations from this relationship above $q \approx 15$ nm^{-1}. This means that the assumption of continuous distribution at small distances (observed by large momentum transfers) is invalid. In the next example we consider discontinuous jump motions explicitly.

EXAMPLE Diffusion of hydrogen in palladium

Diffusion in solids takes place via jumps from one site to another. In general the residence time at a site is long compared to the time the particle needs for the jump. In this case we deal with a noncontinuous diffusion and must take this into account from the beginning.

We discuss a specific diffusion model and then apply the results to hydrogen diffusion in palladium. We assume that hydrogen performs jumps between octahedral sites in a face-centered cubic lattice (see Fig. 10.17).

Denoting the average residence time at a lattice site as $\bar{\tau}$ and assuming completely statistical jumps, the autocorrelation function $G_s(r, t)$ follows the rate equation

$$\frac{\partial G_s(r, t)}{\partial t} = -\frac{1}{\bar{\tau}} G_s(r, t) + \frac{1}{z\bar{\tau}} \sum_{i=1}^{z} G_s(r + l_i, t) \qquad (10.37)$$

The first term on the right side of this equation describes the decrease

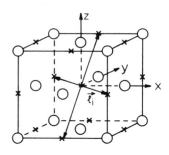

Fig. 10.17 Unit cell of face-centered cubic palladium with octahedral sites denoted by crosses. The coordinate system (x, y, z) used in the text and the direction of the jump vector l_i from the central octahedral site to the 12 nearest-neighbor sites is indicated.

of $G_s(r, t)$ due to jumps out of the site whereas the second term gives the increase due to jumps into the site. The l_i are the space vectors from r to the neighboring sites, and z is the number of nearest neighbors; in the present case $z = 12$. For jumps into the site, the factor $1/z$ appears, since only one of z jumps from any neighboring site leads to the position r.

For the solution of Eq. (10.37), we come back to the function given in Eq. (10.32), the Fourier transform with respect to space. This yields

$$\frac{\partial I_s(q, t)}{\partial t} = -\frac{1}{\bar{\tau}} I_s(q, t) + \frac{1}{z\bar{\tau}} \sum_{i=1}^{z} \int G_s(r + l_i, t) \exp(iq \cdot r) d^3r$$

$$= -\frac{1}{\bar{\tau}} I_s(q, t) + \frac{1}{z\bar{\tau}} \sum_{i=1}^{z} \exp(iq \cdot r) I_s(q, t)$$

$$= -\frac{1}{\bar{\tau}} f(q) I_s(q, t) \tag{10.38}$$

with

$$f(q) = \frac{1}{z} \sum_{i=1}^{z} [1 - \exp(iq \cdot l_i)] \tag{10.39}$$

The solution of Eq. (10.38) gives

$$I_s(q, t) = \exp\left[-\frac{f(q)}{\bar{\tau}} t\right] \tag{10.40}$$

Fourier transformation with respect to time yields the scattering function

$$S_{\text{inc}}(q, \omega) = \frac{1}{\pi} \frac{f(q)/\bar{\tau}}{\omega^2 + [f(q)/\bar{\tau}]^2} \tag{10.41}$$

Thus one again obtains (see Eq. (10.35)) a Lorentz distribution. The width is

$$\Delta \omega = \frac{2f(q)}{\bar{\tau}} \tag{10.42}$$

In contrast to Eq. (10.36), the width depends on the direction of q, not just on the magnitude of q. This will become even clearer if $f(q)$ is introduced in the present example (Fig. 10.17). This yields

$$f(q) = \frac{1}{6}\left\{6 - \cos\left[\frac{a}{2}(q_x + q_y)\right] - \cos\left[\frac{a}{2}(q_x - q_y)\right]\right.$$
$$- \cos\left[\frac{a}{2}(q_x + q_z)\right] - \cos\left[\frac{a}{2}(q_x - q_z)\right]$$
$$\left. - \cos\left[\frac{a}{2}(q_y + q_z)\right] - \cos\left[\frac{a}{2}(q_y - q_z)\right]\right\} \quad (10.43)$$

where a is the lattice constant of the face-centered cubic lattice and q_x, q_y, and q_z are the components of the scattering vectors in Cartesian coordinates along the cube axes. Assuming a different jump model, e.g. jumps among tetrahedral sites, would yield a completely different function for $f(q)$, and one would obtain a different dependence of the linewidth on q. Thus, one can determine the jump geometry from the dependence of the quasi-elastic width on q. This is demonstrated in Fig. 10.18.

The linewidths of quasi-elastic neutron scattering for hydrogen in palladium are shown in Fig. 10.18. The scattering vector q is parallel to the $\langle 100 \rangle$ direction, i.e. $q_x = q$ and $q_y = q_z = 0$, on the left half of the

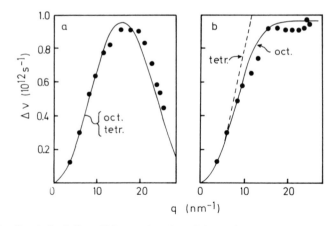

Fig. 10.18 Quasi-elastic linewidths as a function of the magnitude of the scattering vector q for (a) q parallel to the $\langle 100 \rangle$ direction and (b) q parallel to the $\langle 110 \rangle$ direction. The solid lines are calculated for jumps between neighboring octahedral interstitial sites with a mean residence time at the lattice site of $\bar{\tau} = 2.8 \times 10^{-12}$ s. The dashed lines were calculated assuming jumps among tetrahedral sites. For scattering with q parallel to the $\langle 100 \rangle$ direction, both curves coincide and are in fair agreement with the experimental data. The calculated curves are clearly different for the $\langle 110 \rangle$ direction, and only the assumption of octahedral positions agrees with the data (Rowe et al., 1972).

figure. On the right side, q is parallel to the $\langle 110 \rangle$ direction, i.e. $q_x = q_y = q/\sqrt{2}$ and $q_z = 0$. The solid lines are calculated using Eqs. (10.42) and (10.43). Time $\bar{\tau}$ is taken as an adjustable parameter, and its value, determined by fitting to the experimental data, is 2.8×10^{-12} s. The dashed lines in Fig. 10.18 are the best-fit calculations for the tetrahedral site model. These do not agree with the data, and this model is discarded.

The good agreement between the measured and calculated data (for the octahedral site) found here is more the exception than the rule. Similar experiments for hydrogen in V, Nb, and Ta show strong deviations from calculations for simple nearest-neighbor jump models. Jumps to more-distant sites are assumed to play a role in these systems.

10.6 Inelastic Neutron Scattering

In many cases, inelastic neutron scattering is an ideal method for studying elementary excitations in solids (phonons, magnons, etc.). It is applicable if the neutron absorption by nuclei in the material is not too large. In inelastic neutron scattering, a phonon or magnon with wave vector q and energy $\hbar\omega(q)$ is created or annihilated. The Bragg relation and energy conservation requirements give

$$k = k' + G_{hkl} \pm q$$
$$\frac{\hbar^2 k^2}{2m_n} = \frac{\hbar^2 k'^2}{2m_n} \pm \hbar\omega(q) \tag{10.44}$$

with G_{hkl} the reciprocal lattice vector, m_n the mass of the neutron, and k and k' the wave vectors of the incoming and scattered neutrons respectively. The plus sign applies for creation and the minus sign for annihilation of an elementary excitation.

EXAMPLE Phonon dispersion in copper

Copper has a face-centered cubic crystal structure and has only one atom in the primitive unit cell. In a very simple vibration model one assumes that the atomic planes vibrate as a whole. In this simple model, the dispersion relation is

$$\omega(q) = \sqrt{\frac{4C}{M}} \sin\left(\frac{1}{2}qd\right) \tag{10.45}$$

where C is the spring constant, M the mass of the atoms, and d the equilibrium separation of the planes. For q parallel to the $\langle 100 \rangle$ direction $d = a/2$ (a is the lattice constant), and one obtains the function shown in Fig. 10.19.

The spring constants for longitudinal and transverse waves are different, so the phonon dispersion curve splits into different branches. The experimental values for ω in three different directions are shown in Fig. 10.20. The solid lines were obtained by realistic model calculations.

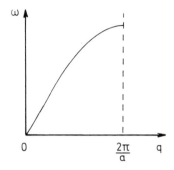

Fig. 10.19 Dispersion relation for the simple model of vibrating planes in face-centered cubic crystals with q parallel to $\langle 100 \rangle$.

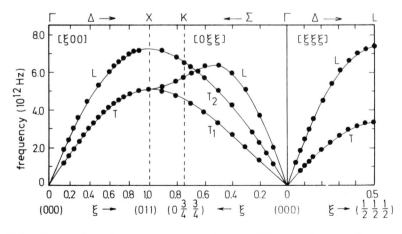

Fig. 10.20 Phonon dispersion curves for copper in three different q directions (Dorner, 1982). Symbols for the symmetry points in reciprocal space are shown at the top of the figure. The solid lines were obtained by model calculations (Nicklow et al., 1967). One sees that the sinusoidal curve of Fig. 10.19 reproduces the actual behavior only roughly.

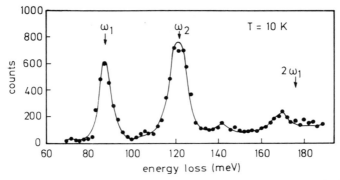

Fig. 10.21 Energy loss spectrum for inelastically scattered neutrons in a niobium deuteride ($NbD_{0.85}$) sample at 10 K (Richter and Shapiro, 1980).

EXAMPLE Local vibrations of hydrogen isotopes in niobium

Hydrogen occupies tetrahedral sites in the metals Nb and Ta, which have a body-centered cubic structure. The vibrational behavior of hydrogen at this site can be roughly described by the Einstein model. In this model, hydrogen is assumed to be in a fixed potential produced by the surrounding ions and electrons (local modes). It is implicitly assumed that the vibrations of hydrogen do not influence the potential.

In a bcc lattice, the tetrahedral interstitial sites are not perfect tetrahedra but are somewhat distorted. Consequently the potential at the tetrahedral site is not isotropic but is somewhat less curved along the tetragonal axis than in the two directions perpendicular to this axis. Thus, two different oscillator frequencies are expected, ω_1 and ω_2, with the intensity of ω_2 being twice that of ω_1. In addition, the relative magnitude of ω_1 and ω_2 can be predicted. If one assumes that hydrogen is bound by longitudinal spring forces to the four atoms forming the tetrahedron, one expects $\omega_2/\omega_1 = \sqrt{2}$ (Richter and Shapiro, 1980).

Fig. 10.21 shows the result of inelastic neutron scattering on niobium deuteride (Richter and Shapiro, 1980). We have chosen this example (and not NbH) because very impressive data are available for the deuteride. Frequencies ω_1 and ω_2 designate the frequencies of local vibrational modes parallel and perpendicular to the tetragonal axis respectively. The energies of these lines and the intensity ratio agree approximately with the model discussed above.

An additional line is seen in the spectrum at an energy loss of 170 meV. This line corresponds to the excitation of two vibration quanta. If the potential were completely harmonic, the position of this line would be expected at somewhat higher energy (see arrow in Fig. 10.21 at $2\omega_1$). The anharmonicity of the potential can be derived from the difference between the actual and expected line position.

11
Ion Beam Analysis

Ion beams are very important for materials analysis, i.e. for the determination of the elemental composition of materials and, in some instances, for determination of atomic arrangement. The various methods of ion beam analysis are based on either the Coulomb interaction of ion beams with the constituents of the solid or nuclear reactions induced by energetic ions on nuclei in the substance. In this chapter we will primarily discuss three methods of ion beam analysis:

(a) Rutherford scattering (Rutherford backscattering (RBS) and elastic recoil detection analysis (ERDA))
(b) Channeling
(c) Nuclear reaction analysis (NRA)

Other ion beam methods include proton-induced X-ray emission (PIXE) and charged particle activation analysis (CPAA). These methods will be mentioned only briefly here.

In PIXE, protons with energy 1 to 2 MeV are directed onto the sample, causing electrons from inner shells (K and L shells) to be ejected. When these inner shells are filled, characteristic X-rays are emitted and are used for identification of the elements in the sample. The intensity of the characteristic X-rays is proportional to the concentration of the element. An advantage of PIXE over electron-induced X-ray emission is that the bremsstrahlung background is lower and therefore higher sensitivity is attained.

In charged particle activation analysis (CPAA) a radioactive nuclide is produced by a nuclear reaction (e.g. $^{12}C(^{3}He, \alpha)^{11}C$) which decays with a known lifetime ($t_{1/2}(^{11}C) = 20.3$ min). The radioactive nucleus can be identified by measuring the lifetime, and the induced activity gives the concentration of the target isotope (here ^{12}C). The charged particle activation analysis is in competition with neutron activation analysis, which is used mainly for heavy elements. For light elements there is often no appropriate nuclear reaction that can be induced by neutrons, so CPAA must be used.

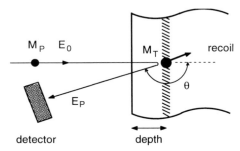

Fig. 11.1 Block diagram of Rutherford backscattering.

11.1 Rutherford Backscattering (RBS)

Rutherford backscattering uses elastic backscattering of charged particles by nuclei in the solid. In this method the collision kinematics is exploited; since the energy of the scattered projectile depends on the mass of the scatterer, the atoms in the solid can be identified. A block diagram of the Rutherford backscattering principles is shown in Fig. 11.1.

In RBS, the important quantities are: the energy transfer in the collision (kinematic factor), the cross section of the reaction, and the energy loss of the incoming and outgoing particles. These properties are discussed next.

11.1.1 Kinematic Factor

In the collision of two nuclei, the interaction is Coulombic as long as the energy is small enough that nuclear forces do not play a role. The ratio of the projectile energy E_P after the scattering to the energy E_0 before scattering defines the so-called kinematic factor,

$$K := \frac{E_P}{E_0} \tag{11.1}$$

where index P is used for the projectile, and T will be used for the target atom. Energy and momentum conservation lead to

$$K = \left[\frac{\sqrt{1 - [(M_P/M_T)\sin\theta]^2} + (M_P/M_T)\cos\theta}{1 + (M_P/M_T)} \right]^2 \tag{11.2}$$

for the kinematic factor K in an elastic collision. As can be seen, the kinematic factor depends only on the mass ratio M_P/M_T and the scattering angle θ.

We will now discuss what angles are most favorable for RBS measurements. For a given ratio M_P/M_T and a fixed incident energy E_0, K should be as small as possible, since then the projectile identification, via energy loss, is optimum. This is of course fulfilled at backscattering ($\theta = 180°$), since then the projectile transfers maximum energy to the target atom. One then has

$$K(\theta = 180°) = \left[\frac{1 - (M_P/M_T)}{1 + (M_P/M_T)}\right]^2 \quad (11.3)$$

However, this is possible only if $M_P < M_T$, since otherwise no backscattering is possible. For an illustration, the behavior of K as a function of the mass ratio M_T/M_P is shown in Fig. 11.2 as a function of scattering angle.

For optimum distinction between two atomic species with masses M_T and $M_T + \Delta M_T$ in the target, the scattering angle θ has to be chosen such that the mass difference ΔM_T can be measured most accurately. As can be seen from the slopes of the curves in Fig. 11.2, the maximum rate of change of K as a function of M_T occurs at a scattering angle of 180° except for small mass ratios ($M_P/M_T \leq 3$).

Thus the most favorable detector position is at $\theta = 180°$; therefore the method is called Rutherford *back*scattering. In practice, however, angles of order $\theta = 170°$ are the largest that can be attained, since the incident particle beam cannot pass through the detector. An efficient arrangement can be obtained by using an annular detector. The dependence of the kinematic factors for the angle $\theta = 170°$ as a function of target mass M_T is shown in Fig. 11.3 for different projectiles.

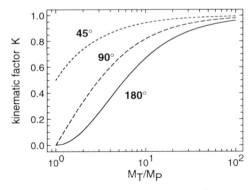

Fig. 11.2 Kinematic factor K as a function of the mass ratio M_T/M_P for different scattering angles.

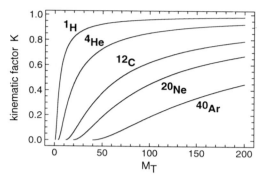

Fig. 11.3 Kinematic factor for $\theta = 170°$ as a function of the target mass M_T for different projectiles. For $M_T < M_P$, no backscattering occurs.

11.1.2 Cross Section for Rutherford Backscattering

In the application of the Rutherford backscattering method for detecting atomic species in the sample, it is important to know the probability of the scattering process. This probability W for a reaction is determined by the cross section σ and the number of scattering centers N per irradiated target area A,

$$W = \frac{N}{A}\sigma \tag{11.4}$$

Fig. 11.4 illustrates this behavior. A schematic illustration of the quantity σ is obtained if one assumes that particles hitting the atom in an area σ are scattered whereas all others pass completely unaffected. The quantity σ is then the effective area in which scattering occurs. Eq. (11.4) is of course valid only if the target is so thin that these effective areas do not overlap. The number of scattering events per time is then

$$I = I_0 W = \frac{I_0 N}{A}\sigma \tag{11.5}$$

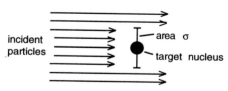

Fig. 11.4 Schematic illustration of the cross section.

where I_0 is the number of incident particles per unit time. The differential cross section $d\sigma/d\Omega$ is defined in analogy to Eq. (11.5) as

$$\frac{d\sigma}{d\Omega} = \frac{A}{I_0 N} \frac{dI}{d\Omega} \qquad (11.6)$$

where $dI/d\Omega$ describes the scattering events per unit time observed in the solid angle element $d\Omega$.

In Rutherford scattering, pure Coulomb interaction between the two colliding nuclei is assumed. Since Coulomb interaction has infinite range, the total cross section for Rutherford scattering would be infinite. In reality, the cross section is finite, since the nucleus, with its electronic shell, appears neutral from outside the atom. The experimentally detected backscattering, however, occurs only in close encounters; there the shielding effects can generally be neglected, and the assumption of Coulomb forces is valid. Deviations from pure Coulomb scattering may occur also if the energy of the particles is so large that the Coulomb barrier of the nucleus can be surmounted, and nuclear forces become effective. In the following discussion, we limit attention to pure Coulombic forces.

The differential cross section for Rutherford scattering in the laboratory system is

$$\frac{d\sigma}{d\Omega} = \left(\frac{Z_P Z_T e^2}{16\pi\varepsilon_0 E}\right) \frac{4}{\sin^4 \theta} \frac{[\sqrt{1 - [(M_P/M_T)\sin\theta]^2} + \cos\theta]^2}{\sqrt{1 - [(M_P/M_T)\sin\theta]^2}} \qquad (11.7)$$

with Z_P and Z_T the atomic numbers of the projectile and target atom respectively. For the case where $M_P \ll M_T$, Eq. (11.7) transforms into the well-known (center-of-mass) Rutherford formula

$$\frac{d\sigma}{d\Omega} = \left(\frac{Z_P Z_T e^2}{16\pi\varepsilon_0 E}\right)^2 \frac{1}{\sin^4(\theta/2)} \qquad (11.8)$$

Two important consequences can be seen from the formula for the differential cross section. First, the cross section is proportional to $(Z_P Z_T)^2$, i.e. for a given target atom it is favorable to use projectiles with higher atomic number (^{4}He instead of ^{1}H). Secondly, the cross section is proportional to E^{-2}, which means that the scattering rate decreases substantially with higher incident energy.

11.1.3 Energy Loss in Matter

Charged particles lose energy when passing through matter. Except for very low energies, scattering of electrons, i.e. excitation or ionization of target atoms, is the dominant process for the energy loss (electronic energy loss). Scattering of the projectile by target nuclei is much less frequent; for protons, deuterons, alpha-particles, and heavy ions, the nuclear energy loss is negligible for energies above 100 keV/nucleon. Multiple scattering causes energy and angular straggling. The energy loss per unit length (stopping power) at energy E of the incident beam is defined as

$$-\frac{dE}{dx} := -\lim_{\Delta x \to 0} \frac{\Delta E}{\Delta x} \quad (11.9)$$

Bethe and Bloch have calculated electronic energy losses; they find the following,

$$-\frac{dE}{dx} = n \frac{Z_P^2 Z_T e^4}{4\pi\varepsilon_0^2 v^2 m_e} \left[\ln \frac{2m_e v^2}{\langle I \rangle} - \ln\left(1 - \frac{v^2}{c^2}\right) - \frac{v^2}{c^2} \right] \quad (11.10)$$

where Z_P is the atomic number, v the velocity of the incident particle, and n is the number density of target atoms with atomic number Z_T. The quantity $\langle I \rangle$ is the average ionization potential and is estimated to be $\langle I \rangle = 11.5 Z_T$ (eV).

Since the projectile velocities are very small compared to the velocity of light c for ion beam techniques, the last two terms in Eq. (11.10) can be neglected.

In Fig. 11.5 the experimentally determined stopping power for ^{4}He projectiles in silicon is shown along with the stopping power calculated using

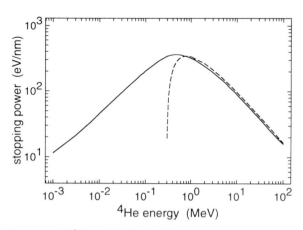

Fig. 11.5 Stopping power of ^{4}He projectiles in silicon as a function of the projectile energy. The solid line represents the experimental data (Ziegler, 1977). The dashed line was calculated from the Bethe–Bloch formula (Eq. (11.10)).

the Bethe–Bloch formula. The dependence can be understood qualitatively in the following way. In the energy range $\langle I \rangle \ll E \ll M_P c^2$, the first logarithmic term in the parenthesis of Eq. (11.10) varies slowly and the relativistic terms can be neglected, so that one obtains approximately

$$-\frac{dE}{dx} \propto \frac{1}{v^2} \propto \frac{1}{E} \tag{11.11}$$

The first logarithmic term becomes important for small energies, and the curve decreases steeply below $E \approx 500 \langle I \rangle$. At lower energies, the interaction of the projectile with weakly bound electrons causes additional energy loss. In this region the nuclear energy loss also becomes relevant. Both effects lead to deviations from the Bethe–Bloch formula.

The discussion of the stopping power thus far has been confined to substances with only a single element. For matter which is composed of several elements, one assumes additivity of the stopping power normalized to the atomic densities (Bragg–Kleeman rule). The stopping power of a substance composed of elements A and B with relative abundances α and β is then

$$\frac{dE}{dx}(A_\alpha B_\beta) = n(A_\alpha B_\beta)\left[\frac{\alpha}{n(A)}\frac{dE}{dx}(A) + \frac{\beta}{n(B)}\frac{dE}{dx}(B)\right] \tag{11.12}$$

The specific energy loss for various projectile–target atom combinations is tabulated in Ziegler (1977). Approximate formulas for the stopping power can also be found in this reference.

11.1.4 Acceleration and Detection of Charged Particles

Particle accelerators

For ion beam analysis, particle beams of several MeV are needed. Electrostatic accelerators are commonly used to produce these beams. The underlying principle is that charged particles fall through a potential energy difference U and gain kinetic energy

$$E = qeU \tag{11.13}$$

where qe is the charge of the particle. Other acceleration principles are based on alternating electric fields where the phase of the alternating field with respect to the particle path is such that on average the particle is accelerated (the particle rides on a wave). This alternating field principle is used in cyclotrons and synchrotrons as well as in high-frequency linear accelerators; generally, with the alternating field principle higher energies can be obtained than by electrostatic acceleration. The higher energies are needed in nuclear and elementary particle physics but generally not in ion beam analysis. Therefore we limit our discussion to electrostatic acceleration.

The main components of an electrostatic accelerator are the ion source, the acceleration tube, and the device for producing the accelerating voltage.

An example of an ion source is the gas discharge tube in which a plasma is produced by electron impact. The ions are extracted from the plasma by an extraction voltage and are guided into the evacuated acceleration tube. Materials which are not available in gaseous compounds can be evaporated in an oven and then transferred by a carrier gas (e.g. Ar) into the discharge area.

Only one more of the many different types of ion sources is discussed here — the so-called sputter source in which ions are directly ejected from a solid by ion bombardment. In addition to neutral atoms and positive ions, some fraction of negative ions are also created in this process. Negative ions are necessary if tandem accelerators (see below) are used.

Electrostatic accelerators use either the Cockcroft–Walton or Van de Graaff principle. The Cockcroft–Walton method, which is not discussed in more detail here, uses voltage multiplication produced by an alternating current which flows through a multiplying and rectifying circuit.

A schematic illustration of a Van de Graaff-type accelerator is shown in Fig. 11.6. This type of accelerator was developed in the 1920s. This accelerator uses a rotating belt with which charge is transported mechanically into the high-voltage region. The charges are sprayed onto the insulating belt at ground potential, transported into the high-energy region, and then removed there. The high-energy side is enclosed in an electrically conducting sphere in whose inner part a field-free region exists so that the removal of the charges can take place. The ion source is also placed in this

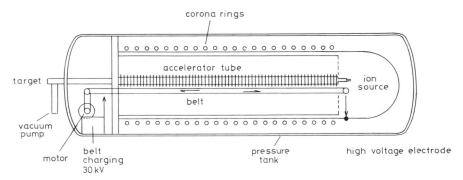

Fig. 11.6 Block diagram of a Van de Graaff accelerator.

high-energy terminal. Instead of the belt, nowadays, chains with insulated links are often used.

The acceleration tube (see Fig. 11.6), which is evacuated by a vacuum pump, consists of equidistant metal rings which are separated from one another by glass cylinders. The glass and metal parts are glued together and form a vacuum-tight tube. In order to obtain a homogeneous voltage drop over the whole length of the acceleration tube, the metal rings of the acceleration tube are connected with corona rings which surround the tube on the outside. The corona rings are connected through a resistance chain to provide the homogeneous voltage drop.

Finally we should mention the high-pressure tank which is filled with a quench gas to prevent high-voltage breakdown. In air, voltages up to only about 500 keV can be maintained; with a quench gas (e.g. SF_6 at 10 bar) voltages of 10 MV and higher are obtainable.

In a tandem accelerator, the high voltage is used twice. The high-voltage part is placed in the middle of an accelerator. In the first stage, negative ions (the ion source is at ground potential) are accelerated towards the positively charged high-energy terminal. There their charge is changed from negative to positive by allowing the particles to pass through a foil or gas cell. Afterwards the positive ions are accelerated further to the other end of the accelerator. The total energy achieved is

$$E = (q + 1)eU \qquad (11.14)$$

where q is the charge state of the ion after the charge exchange. The 1 in Eq. (11.14) accounts for the acceleration of the (singly charged) negative ion in the first stage of the accelerator. For heavy ions q can assume values much larger than 1, so that the energy can be not just doubled but multiplied by several factors.

Particle detectors

For detection of charged particles (protons, α-particles, etc.) semiconductor detectors are often used. The most common is the Si surface barrier detector, which will be discussed in some detail here.

Semiconductor detectors (see Ge detector in Section 2.4) work on a principle similar to an ionization chamber. The energetic charged particle passing through the semiconductor creates electron–hole pairs along its path (in the ionization chamber electrons and ions) which are separated by an electric field and create a current pulse. The amplitude of the current pulse is proportional to the number of electron–hole pairs and therefore to the energy of the ionizing particle. The creation of an electron–hole pair in silicon requires an average energy of 3.7 eV. Thus an α-particle of 5 MeV energy produces approximately

$$\frac{5 \times 10^6 \text{ eV}}{3.7 \text{ eV/pair}} \approx 1.7 \times 10^6 \text{ electron–hole pairs}$$

in Si. In an ionization chamber the creation of an electron–ion pair requires approximately 30 eV. Thus, for the same energy of the passing particle, approximately one tenth as many charge carriers are produced as in semiconductors. The larger number of carriers and consequent lower statistical energy uncertainty gives semiconductor detectors an advantage over ionization chambers. In addition, the semiconductor detectors have better time resolution, because the charge is collected more rapidly at the semiconductor electrodes.

A practical semiconductor detector uses a p-n junction. For a detailed description of p-n junctions, the reader is referred to standard texts in solid state physics. Here we summarize only essential points (see Fig. 11.7).

If p-type and n-type semiconducting materials are brought into contact, an exchange of charge carriers occurs. The electrons from the n-type material flow into the p-type material, and the positively charged holes from the p-type material flow into the n-type side. Therefore a voltage difference appears, and further charge flow is inhibited. In the interface region between p- and n-type materials, there is a depletion region. Application of an external reverse-bias electric voltage (positive electrode on the n-type

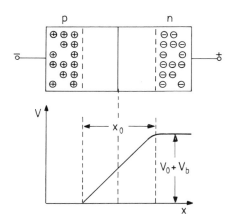

Fig. 11.7 A p-n junction in a semiconductor with applied voltage V_0 in reverse-bias direction. V_b is the 'built-in' voltage of the p-n junction. A carrier depletion zone having width x_0 exists between the p-type (excess of mobile holes) and the n-type (excess of mobile electrons) regions. The '$\oplus$' and '$\ominus$' represent mobile carriers. The voltage as a function of distance x from the surface is shown in the lower part of the figure.

and negative electrode on the p-type side) increases the voltage difference and broadens the depletion region.

To make a Si surface barrier detector one starts with a thin slice of weakly n-type silicon that is slightly oxidized on one surface. The oxidation causes surface states which have p-type character. Thus one has a p-type part of a diode on the n-type base material. A thin gold film is evaporated on the p-type front side and an aluminum film on the back side to make contacts and protect the diode.

An alternative to the surface barrier detector is the so-called ion-implanted detector. In this case, the p-type layer is produced by boron implantation into n-type base material. This fabrication is somewhat more complicated but has the advantage that the surface of the detector is mechanically more stable than in the usual surface barrier detector. Another possibility is to use p-type instead of n-type base material and produce an n-type surface layer. The differences are important in practice, but the operating principles are the same.

The depletion region (see Fig. 11.8) is of decisive importance to proper functioning of the device, since only the electron–hole pairs produced in this region contribute to the current pulse. Thus the layer on the front of the detector (gold contact and p-type layer) should be thin, since the ionization in this part does not contribute to the current pulse. In addition, the depletion zone should be thick enough to stop the particle completely in order that the total energy contributes to the pulse. The width of the depletion zone is (Sze, 1985)

$$x_0 = \sqrt{\frac{2\varepsilon(V_0 + V_b)}{qn_B}} \tag{11.15}$$

where ε is the dielectric constant, V_b the built-in p-n voltage, V_0 the external

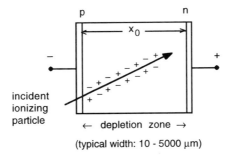

(typical width: 10 - 5000 μm)

Fig. 11.8 The incident ionizing particle creates electron–hole pairs along its track in the semiconductor. The charges are separated in the electric field and produce a current pulse in the electrodes.

reverse-bias voltage, q the charge, and n_B the concentration of donor atoms in the n-type base material. One sees in Eq. (11.15) that, by reducing the base-material donor doping concentration or by increasing the bias voltage, the width of the depletion zone can be increased. The base-material doping level is the limiting factor in producing weakly doped high-purity silicon. The maximum bias voltage is limited by breakdown of the diode.

An energy resolution of approximately 15 keV is achievable for 5 MeV α-particles using a surface barrier detector. Fig. 11.9 shows a portion of the energy spectrum for a ^{241}Am calibration source. One sees that the α-lines at 5.433 MeV and 5.477 MeV are well-separated.

In some cases it is important to determine the particle species as well as its energy. This can be done with a detector telescope that consists of two detectors placed one behind the other. The first detector must be so thin that the charged particle can pass through and deposit only part of its energy (ΔE detector). In the subsequent thick detector, the particle is stopped and deposits the rest of its energy (E detector).

The operating principle of the telescope can easily be understood for particles above the maximum of the energy loss curve (see Fig. 11.5), i.e. in the region where the Bethe–Bloch formula (Eq. (11.10)) holds. In this region dE/dx is proportional to $1/E$ so that the product of the energy loss ΔE in the thin detector and the total energy of the particle ($\Delta E + E_{rest}$), where E_{rest} is the energy detected in the second detector, is

$$\Delta E \times (\Delta E + E_{rest}) \propto Z_P^2 M_P \qquad (11.16)$$

where Z_P is the atomic number and M_P the mass of the ionizing particle. For protons $Z_P^2 M_P = 1$ (M_P in atomic units) whereas for α-particles

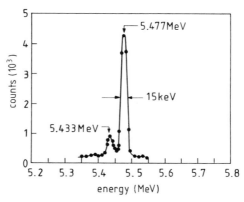

Fig. 11.9 Energy spectrum of a ^{241}Am α-source measured with a surface barrier detector.

$Z_P^2 M_P = 16$. This shows that the two particle species can be well-distinguished by a $\Delta E - E$ telescope. Particles can also be discriminated below the maximum of the energy loss, but the simple relation in Eq. (11.16) is inadequate. More elaborate particle discrimination (usually with the aid of computers) must be made.

In addition to semiconductor detectors, ionization chambers are used for detection of charged particles despite the disadvantages mentioned above. Ionization chambers are sometimes preferable for heavy-ion detection, because heavy ions cause substantial radiation damage in semiconductors and therefore reduce their useful lifetime. Ionization chambers are insensitive to radiation damage; in addition, large solid angles can be realized.

11.1.5 Experiments on Thin Films

An important field of application for Rutherford backscattering is the analysis of thin-film and multilayer materials. This application will be discussed as an illustration to clarify the basic concept of the Rutherford backscattering method.

First we will discuss the relationship between the measured energy of the scattered particle and the depth from which it was scattered. We will consider the energy of a scattered particle if it is scattered either at the surface or at a depth x (see Fig. 11.10).

For calculation of the energy E_P after the scattering at depth x in the material, we assume that the stopping power dE/dx on the way in and on the way out is constant (this is approximately correct if the total energy loss is small compared to the total energy) but in general is different before and after the collision. For the energy E_x and E_P, one then has

$$E_x = E_0 - \left.\frac{dE}{dx}\right|_{E_0} \frac{x}{\sin \alpha_1} \qquad (11.17)$$

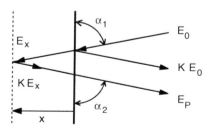

Fig. 11.10 Energy of particles scattered from atoms at the surface or from atoms at depth x. The particle paths form respective angles α_1 and α_2 with respect to the surface.

and

$$E_P = KE_x - \left.\frac{dE}{dx}\right|_{KE_x} \frac{x}{\sin \alpha_2} \qquad (11.18)$$

Using these two equations, the energy difference ΔE between scattering at the surface and scattering at depth x can be calculated

$$\Delta E = KE_0 - E_P = \left[\frac{K}{\sin \alpha_1} \left.\frac{dE}{dx}\right|_{E_0} + \frac{1}{\sin \alpha_2} \left.\frac{dE}{dx}\right|_{KE_x}\right] x \qquad (11.19)$$

If x is not known, we also do not know E_x, which is required for the calculation of the energy loss (see Eq. (11.19)). To solve this problem we make use of the relatively weak dependence of the energy loss on energy and set

$$\left.\frac{dE}{dx}\right|_{KE_x} = \left.\frac{dE}{dx}\right|_{KE_0} \qquad (11.20)$$

This yields

$$\Delta E = \left[\frac{K}{\sin \alpha_1} \left.\frac{dE}{dx}\right|_{E_0} + \frac{1}{\sin \alpha_2} \left.\frac{dE}{dx}\right|_{KE_0}\right] x \qquad (11.21)$$

Thus one obtains a linear relation between the energy of the backscattered particle and the depth at which the scattering took place.

These considerations will now be applied to a thin self-supporting film which consists only of one element. We limit ourselves to perpendicular incidence of the projectile ($\alpha_1 = 90°$) and exact backscattering ($\alpha_2 = 90°$, $\theta = \alpha_1 + \alpha_2 = 180°$). The shape of the spectrum and the physical quantities that can be derived from it will be discussed. A schematic backscattering spectrum for a thin film is shown in Fig. 11.11.

A realistic backscattering spectrum consists of a histogram in which the number of backscattered particles $H(i)$ in the energy interval Δ is accumulated at the corresponding energies $E_P(i)$ where i is a channel index. The bin size of the histogram is limited by the energy resolution of the particle detector. Measuring points in an energy range Δ correspond to scattering events in the film from a layer with thickness Δx.

As can be seen qualitatively from Fig. 11.11, the layer thickness d can be directly obtained from the width of the energy distribution ΔE of the scattered particles. According to Eq. (11.21), one has

$$\Delta E = \left[K \left.\frac{dE}{dx}\right|_{E_0} + \left.\frac{dE}{dx}\right|_{KE_0}\right] d \qquad (11.22)$$

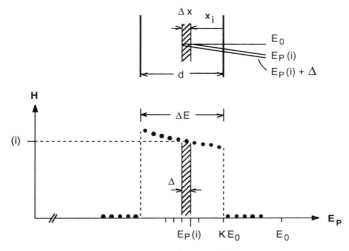

Fig. 11.11 Rutherford backscattering spectrum (schematic) for a thin self-supporting film.

Thus, the specific energy loss of the particle beam in the film must be known in order to determine the layer thickness. The kinematic factor K can be taken directly from the position of the upper edge of the energy distribution (see Fig. 11.11).

The number of backscattered particles $H(i)$ in the ith energy interval can be calculated in the following way (compare Eq. (11.6))

$$H(i) = \frac{dI}{d\Omega}\Delta\Omega t_0 = \left.\frac{d\sigma}{d\Omega}\right|_{E(x_i)} I_0 \frac{N_i}{A} \Delta\Omega t_0 \qquad (11.23)$$

where $\Delta\Omega$ is the solid angle of the detector, and t_0 the measuring time. N_i/A is the number of target atoms per irradiated area A in the layer Δx corresponding to the energy interval Δ. Using the relation $N_i/A = n\Delta x$, with $n = N/V$ (density of target atoms in the irradiated volume V), one obtains

$$H(i) = \left.\frac{d\sigma}{d\Omega}\right|_{E(x_i)} I_0 \Delta\Omega t_0 n \Delta x \qquad (11.24)$$

We can now calculate the area F under the backscattering spectrum

$$F = \sum_i H(i) = I_0 \Delta\Omega t_0 n \sum_i \left(\left.\frac{d\sigma}{d\Omega}\right|_{E(x_i)} \Delta x\right) \qquad (11.25)$$

For the case that the cross section is independent of depth x, i.e. if the recoil spectrum is rectangular, one obtains the following simple result

$$F = I_0 \Delta\Omega \, t_0 nd \left.\frac{d\sigma}{d\Omega}\right|_{E_0} \qquad (11.26)$$

This equation permits one to determine the film thickness d if the target particle density n, the beam intensity, and the detector solid angle are known.

Another useful quantity is the number $H(0)$ of particles scattered from the surface layer. Using Eqs. (11.24) and (11.21), one obtains for $\alpha_1 = \alpha_2 = 90°$

$$H(0) = \left.\frac{d\sigma}{d\Omega}\right|_{E_0} I_0 \Delta\Omega \, t_0 n \Delta \left[K\left.\frac{dE}{dx}\right|_{E_0} + \left.\frac{dE}{dx}\right|_{KE_0}\right]^{-1} \qquad (11.27)$$

One should note that the number of particles backscattered from inside layers increases in general, since the cross section for Rutherford scattering increases with decreasing particle energy (compare Eq. (11.8)). This behavior is indicated in Fig. 11.11.

EXAMPLE Cobalt silicide formation in Co/Si films

The formation of nickel and cobalt silicide compounds and their growth are technologically important for the production of conducting layers in semiconductor devices. Rutherford backscattering has made important contributions to understanding this subject. We will discuss here the work of Langouche and collaborators on this problem (Wu et al., 1991).

A cobalt film was evaporated onto the (111) surface of a silicon single crystal held at room temperature. After preparation a Rutherford backscattering experiment using 2 MeV ^{4}He particles was performed. The resulting spectrum is shown in Fig. 11.12a. One recognizes two different regions: the data points between 1.43 MeV and 1.57 MeV are caused by backscattering from cobalt, whereas the events below 1.1 eV belong to backscattering from silicon ($K_{Co} > K_{Si}$). The cobalt events start, as expected for backscattering from the cobalt surface, at the energy $K_{Co}E_0$ and end, in the present case, at ≈ 1.43 MeV. The events at the low-energy edge of the Co portion of the spectrum correspond to scattering from the back side of the Co film.

The scattering events originating from silicon, however, start not at $K_{Si}E_0$ but at somewhat lower energy. This is due to the fact that the

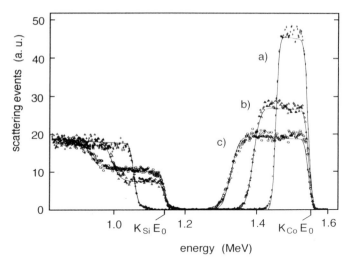

Fig. 11.12 Rutherford backscattering from a Co film on Si(111). (a) Directly after preparation at room temperature. (b) After annealing at 450 °C for one hour, CoSi is formed. (c) After additional annealing at 600 °C for one hour, CoSi$_2$ is formed. After Vantomme (1991).

^{4}He beam has to pass through the Co film and therefore has lost energy when it impinges on the Si surface. The silicon scattering events do not end at lower energy, since we are dealing with a thick silicon substrate. The solid lines are simulations of the recoil spectrum with the theory discussed above for thin films.

If the Co/Si system is annealed and then investigated by Rutherford backscattering at room temperature, clear changes can be seen in the recoil spectrum. Two such spectra are shown in Fig. 11.12b and c for one-hour anneals at temperatures 450 °C and 600 °C, respectively. One observes that scattering events from silicon now appear at $K_{Si}E_0$, meaning that silicon atoms have reacted with the cobalt film and have come to the surface. Secondly, one sees that the recoil distribution which originates from scattering by cobalt extends to lower energies. This means that cobalt has penetrated the silicon substrate. A layer phase containing both cobalt and silicon has clearly formed during annealing.

For the determination of the stoichiometry of the two Co/Si phases, we make use of the total area under the spectrum (see Eq. (11.26)). We determine the area F_{Co} for the cobalt scattering events and the area F_{Si} for those silicon scattering events that stem from the silicon area that has reacted with cobalt. The latter area must be taken in the energy region between $K_{Si}E_0$ and the energy where the scattering events rise to the silicon substrate value. One then obtains

$$\frac{F_{Co}}{F_{Si}} = \frac{n_{Co}}{n_{Si}} \frac{\left.\frac{d\sigma}{d\Omega}\right|_{E_0}^{(Co)}}{\left.\frac{d\sigma}{d\Omega}\right|_{E_0}^{(Si)}} = \frac{n_{Co}}{n_{Si}} \left(\frac{Z_{Co}}{Z_{Si}}\right)^2 \qquad (11.28)$$

The stoichiometry can be determined from the measured areas using this relation. For the case shown in Fig. 11.12b, the analysis yields $n_{Co}/n_{Si} = 1$, indicating that the phase CoSi has formed. For the case shown in Fig. 11.12c, the stoichiometry $CoSi_2$ is found.

For thicker films where the above simplifying approximations do not hold, one can divide the film into slices for numerical analysis and derive the physical parameters in an iterative way.

11.1.6 *Elastic Recoil Detection Analysis (ERDA)*

As shown above, the measured particle in Rutherford backscattering is the scattered projectile. The atom that is knocked out of the target in the collision can also be detected. This alternative method is called elastic recoil detection analysis (ERDA).

The main advantage of the ERDA method over Rutherford backscattering is that the target atom involved in the collision is not only identified by the kinematics but its chemical identity can be measured directly in the detector telescope. In addition, of course, the energy of the particle can be measured, and from that information the particle's original depth in the film can be determined. Since in the ERDA method a large fraction of the projectile's energy has to be transferred to the target atom, this technique requires that the mass of the projectile be comparable to or larger than the mass of the target atom. Therefore heavy-ion projectiles are preferentially used in the ERDA method. The method is particularly well-suited for detection of light elements (from ^1H to ^{16}O) in matrices of heavy elements, whereas Rutherford backscattering is preferentially used for the opposite case, the detection of heavy elements in a light atom matrix. Thus the two techniques are ideally complementary.

A schematic illustration of the ERDA method is shown in Fig. 11.13. A target atom is knocked out of the substance by the incoming projectile, and is generally detected at a small recoil angle ϕ. The depth distribution of the target atom is derived from the measured recoil energy E_T.

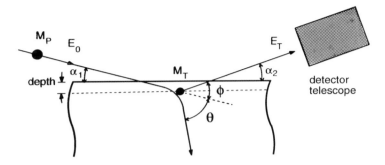

Fig. 11.13 Schematic illustration of the experimental set-up for the ERDA method.

A kinematic factor can once again be defined, but it now refers to the energy E_T of the knocked-out target atom,

$$\widetilde{K} := \frac{E_T}{E_0} \qquad (11.29)$$

where E_0 is the energy of the projectile before scattering. One obtains

$$\widetilde{K} = \frac{4(M_P/M_T)}{[1 + (M_P/M_T)]^2} \cos^2 \phi \qquad (11.30)$$

The recoil angle ϕ is related to the scattering angle θ by

$$\tan \theta = \frac{\sin 2\phi}{(M_P/M_T) - \cos 2\phi} \qquad (11.31)$$

We will demonstrate the application of the ERDA method by discussing an experimental example (Grötzschel et al., 1992). In this experiment a silicon wafer with 150 nm thick boron nitride layer at the surface was investigated. The projectiles used were ^{127}I ions with energy 130 MeV at an incident angle of $\alpha_1 = 15°$. The knocked-out target atom was registered at an angle $\alpha_2 = 15°$ with respect to the probe surface. Thus the total angle for the recoil was $\phi = \alpha_1 + \alpha_2 = 30°$. The detector telescope consisted of an ionization chamber which was divided into a thin front region to measure ΔE and a thick back region to detect the rest of the energy. It is possible to distinguish different atomic species with such a detector telescope.

In Fig. 11.14, the recoil events are shown in a two-dimensional scatter plot with axes ΔE and E_{rest}. The different target atoms can clearly be distinguished in this plot. As can be seen for the Si recoil atoms, the ΔE vs. E_{rot} curve has the behavior shown in Fig. 11.5.

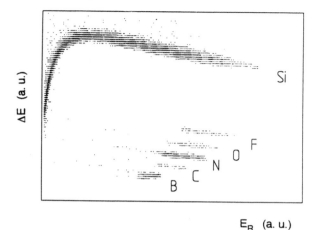

Fig. 11.14 Two-dimensional scatter plot of the recoil events for ^{127}I ion bombardment of a boron nitride film on silicon. The particles were detected with an ionization chamber telescope (Grötzschel et al., 1992).

In Fig. 11.15, the number of recoil events for nitrogen and fluorine (the latter being an impurity in the target) is shown as a function of energy E_T. One recognizes that the profile is that discussed before for Rutherford backscattering on thin films. The layer depth can also be obtained from the energy distribution. In analogy to Eq. (11.21), one has

$$\Delta E = \left[\frac{\widetilde{K}}{\sin \alpha_1} \frac{dE}{dx} \bigg|_{E_0}^{(M_P)} + \frac{1}{\sin \alpha_2} \frac{dE}{dx} \bigg|_{\widetilde{K}E_0}^{(M_T)} \right] d \qquad (11.32)$$

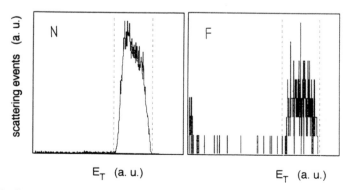

Fig. 11.15 Recoil energy distribution for nitrogen (left) and fluorine (right) from ^{127}I bombardment of a BN film on Si (Grötzschel et al., 1992).

We note that in Eq. (11.32) the stopping power for both the projectile (M_P) and the recoil atom (M_T) appears and that α_2 is the exit angle of the target atom.

The area of the recoil event distribution can easily be calculated. In analogy to Eq. (11.26), one obtains here

$$F = I_0 \Delta \Omega t_0 nd \frac{d\sigma}{d\Omega}\bigg|_{E_0} \quad (11.33)$$

We note that the solid angle element $\Delta\Omega$ in the differential cross section refers now to the recoil angle ϕ. Evaluating the areas of the recoil event distributions for fluorine and nitrogen, one obtains a relative F to N concentration of approximately 10^{-4}.

11.2 Channeling

In the preceding discussion of Rutherford scattering we concentrated on a single collision of an ion with a target atom and considered only fairly large scattering angles. In the following, a phenomenon will be described in which the periodic arrangement of the lattice causes many small correlated glancing collisions. In this way the ion is guided along channels between the atomic rows or atomic planes through small-angle scattering (channeling). In the following, we limit discussion to axial channeling along lattice axes. In this case, channeling occurs if the direction of the incident ion beam coincides with the crystal axis in a single crystal. The path of a channeled projectile is shown schematically in Fig. 11.16. The projectile is guided along the channel and penetrates deeply into the crystal.

To describe channeling one assumes that the charges of the individual nuclei are homogeneously smeared out along the rows. In this picture (continuum model), the projectile scatters on charge strings. Assuming a

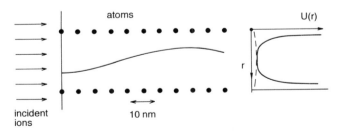

Fig. 11.16 Schematic representation of the path of a channeled ion. The vertical axis is stretched for clarity. The electric potential between the two atomic rows in a continuum model is shown on the right side.

screened Coulomb potential, the potential at distance r in the direction perpendicular to the atomic row is (Feldman and Mayer, 1986)

$$U(r) = \frac{Z_P Z_T e^2}{(4\pi\varepsilon_0)d} \ln\left[\left(\frac{Ca}{r}\right)^2 + 1\right] \quad (11.34)$$

Here d is the average distance between atoms in the row, a is the Thomas–Fermi screening radius, and C is a constant ($C \approx \sqrt{3}$).

The channeling effect depends strongly on the direction of the ion beam with respect to the channeling axis. A criterion for the critical incident angle can be derived using energy conservation. The total energy of the incident projectile in the crystal is given by

$$E = \frac{p_\parallel^2}{2M_P} + \frac{p_\perp^2}{2M_P} + U(r) \quad (11.35)$$

where $p_\parallel$ and $p_\perp$ are the momentum components parallel and perpendicular to the channel direction of the crystal, respectively. These components can be expressed by the angle ψ between the direction of the ion momentum p and the channel direction

$$p_\parallel = p\cos\psi \qquad p_\perp = p\sin\psi \quad (11.36)$$

It follows that the total energy of the projectile can be written as

$$E = \frac{p^2 \cos^2\psi}{2M_P} + \frac{p^2 \sin^2\psi}{2M_P} + U(r) \quad (11.37)$$

Since ψ must be small for channeling to occur, the energy associated with the perpendicular motion is

$$E_\perp \approx \frac{p^2 \psi^2}{2M_P} + U(r) \quad (11.38)$$

This energy is conserved during the passage of the projectile through the channel. The maximum amount of kinetic energy is converted to potential energy at the point of closest approach to the atomic row.

We now can introduce a critical angle ψ_c above which channeling breaks down. This critical angle is derived by assuming that the projectile is not allowed to approach the atomic row closer than a minimum distance $r_{\min}$.

This leads to the relation (see Fig. 11.17)

$$\frac{p^2\psi_c^2}{2M_P} = E\psi_c^2 = U(r_{min}) \tag{11.39}$$

and finally, for the critical angle, to

$$\psi_c = \sqrt{\frac{U(r_{min})}{E}} \tag{11.40}$$

The minimum distance r_{min} is still to be determined. If the projectile approaches an atomic row closer than the average atomic thermal displacement ρ, channeling can be destroyed by collisions with large scattering angle. Using $\langle u^2 \rangle$ for the mean square displacement and assuming isotropic lattice vibrations (compare Section 4.2) we obtain for r_{min}

$$r_{min} \approx \rho = \sqrt{\langle x \rangle^2 + \langle y \rangle^2} = \sqrt{(2/3)\langle u \rangle^2} \tag{11.41}$$

Substituting this condition into Eq. (11.34), one obtains for the critical angle

$$\psi_c = \sqrt{\frac{2Z_P Z_T e^2}{(4\pi\varepsilon_0)dE}} \sqrt{\frac{1}{2}\ln\left[\left(\frac{Ca}{\rho}\right)^2 + 1\right]} \tag{11.42}$$

The Thomas–Fermi shielding radius a has values between 0.01 and 0.02 nm, and the average two-dimensional displacement ρ between 0.005 and 0.01 nm (at room temperature). Therefore the second square root in Eq. (11.42) is of order unity. The experimentally determined critical angles agree reasonably well with this prediction. The temperature dependence is also correctly reproduced.

Even in the case of exact alignment of the ion beam with the channel axis ($\psi = 0°$), there is a non-channeled fraction of ions due to backscattering of projectiles that impinge directly on the atomic row. If the ion beam strikes the atomic row within an area πr_{min}^2 around the atomic row, large-angle

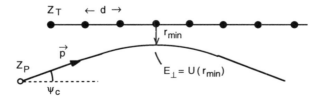

Fig. 11.17 Explanation of the critical angle for channeling.

scattering occurs (Fig. 11.18). The maximum attainable channeling effect, denoted χ_{min}, is thus given by

$$\chi_{min} = \frac{\pi r_{min}^2}{A} = \pi r_{min}^2 nd \tag{11.43}$$

where $A = 1/(nd)$ is the area per atomic row, and n is the atomic density. This leads to typical values of 1–5% for χ_{min}.

For detection of channeling, one uses Rutherford backscattering. The backscattering spectrum is measured as a function of incident angle with respect to the crystal axes (i.e. the tilt angle). For example, the recoil spectrum of ^{1}H ions from nickel is shown on the left side of Fig. 11.19 in (a) a non-channeling and (b) a channeling direction. The scattering events between the dashed lines are summed up and plotted as a function of tilt angle ψ. This gives the typical channeling spectrum shown on the right side of the figure. The critical angle ψ_c discussed above is approximately 1.5° for this example; the minimum backscattering amounts to $\chi_{min} \approx 3\%$.

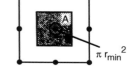

Fig. 11.18 View onto the atomic rows from the ion beam direction.

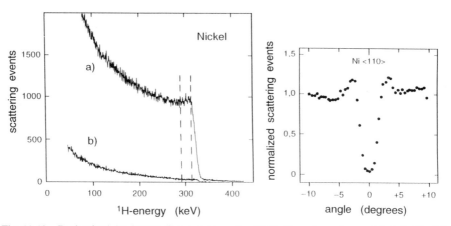

Fig. 11.19 Rutherford backscattering spectrum for 350 keV protons from nickel (left): (a) beam along a random direction, and (b) beam along a ⟨110⟩ direction. The normalized number of scattering events, obtained by summing the events between the dashed lines, is shown on the right side as a function of tilt angle (the channeling spectrum).

11.2.1 Determination of Impurity Lattice Sites in Crystals

An important application of channeling is the determination of lattice sites for impurities in single crystals. For example in semiconductor physics, it is important to know the site occupied by donors and acceptors.

The underlying concept will be illustrated with a two-dimensional lattice. In Fig. 11.20 a lattice is shown in which impurities are incorporated at substitutional and three different interstitial sites. We examine the channeling effect along the crystal axes $\langle 10 \rangle$ and $\langle 11 \rangle$. For the case where the impurities sit in an atomic row with respect to the incident direction and therefore are shadowed by the host atoms, no recoil from the impurities is expected. If on the other hand the impurities are not shadowed by host atoms, an increased backscattering rate is expected for this projectile beam direction, since the channel is now occupied by impurities. The shadowing effect is different for different impurity positions, as can be seen from Fig. 11.20. Thus the position of the impurity in the lattice can be determined by measuring channeling in different directions.

For detection of lattice positions of impurities in single crystals, the recoil events from the impurities must be distinguished from those of the host atoms by the kinematic factor. This is possible for impurities which are heavier than the host atoms. However, very large impurity concentrations ($>0.1\%$) are required since otherwise the measuring time becomes too long, and the channeling method is not applicable.

In the following, an experimental example will be discussed in which these limitations are circumvented. It is a variant of the classical channeling method, namely the so-called emission channeling method (Hofsäss and Lindner, 1991).

In emission channeling one does not work with an external ion beam but

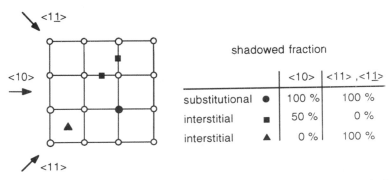

Fig. 11.20 Two-dimensional model of impurities on substitutional and interstitial lattice sites. The expected fraction of blocking is shown on the right side.

incorporates radioactive impurities into the crystal and measures the channeling of the charged decay products (α, e^+, and e^-) leaving the crystal from its interior. The positive charged particles behave as discussed above. The negative electrons, on the other hand, are not guided down the middle of the channel but instead travel close to the atomic rows because of their attractive interaction with the ionic cores.

With the aid of the two-dimensional lattice model (Fig. 11.20), we can easily see how the measured intensities for α-particles depend on the observation direction with respect to the crystal axes. If the radioactive impurities are at substitutional sites (filled circles), the emission of α-particles in directions $\langle 11 \rangle$ and $\langle 10 \rangle$ is blocked by host atoms (blocking effect). On the other hand, if the impurity is on an interstitial site (filled triangle), the α count rate in the direction $\langle 10 \rangle$ is increased, since the particles can move unhindered along this channel. Blocking occurs in the $\langle 11 \rangle$ direction. Impurity positions can be measured by exploiting these characteristic properties of emission channeling.

Hofsäss and collaborators (Wahl et al., 1992) have implanted ^{8}Li isotopes ($t_{1/2} = 843$ ms) into InP single crystals at 60 keV energy using a mass separator. The ^{8}Li isotopes were produced by nuclear reaction in a particle accelerator. The crystal surface was cut to be normal to the $\langle 100 \rangle$ direction. The α-particles ($E_\alpha = 1.3$–5.5 MeV), which are emitted after the β^--decay of ^{8}Li to ^{8}Be, were detected by a two-dimensional position-sensitive silicon detector. The two-dimensional intensity spectrum obtained for the α-particles is shown in Fig. 11.21. One clearly recognizes intensity maxima along

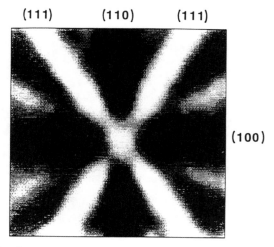

Fig. 11.21 Intensity distribution of α-particles in the two-dimensional α-detector ($T = 283$ K). The center of the picture corresponds to the $\langle 110 \rangle$ emission direction for which axial channeling is observed in this case. Important crystal planes are indicated (Wahl et al., 1992).

the crystal planes (planar channeling). In addition, dark stripes can be recognized which indicate reduced α-emission (blocking effect).

Plotting the α intensity as a function of the emission angle with respect to a crystal axis, one obtains the usual channeling spectra. These are shown in Fig. 11.22 for the $\langle 100 \rangle$ and $\langle 110 \rangle$ directions. One sees that the intensity at an implantation temperature of 300 K is increased in the $\langle 110 \rangle$ direction, and is reduced in the $\langle 100 \rangle$ direction. Emission in the $\langle 111 \rangle$ direction is also reduced, but the spectrum is not shown. This result shows that the ^{8}Li impurities are located at tetrahedral interstitial sites in the zincblende structure of InP. The channeling results are illustrated in the lattice model shown in Fig. 11.22.

The channeling spectra for an implantation temperature of 450 K show reduced emission in all three directions ($\langle 100 \rangle$, $\langle 110 \rangle$, and $\langle 111 \rangle$). A blocking effect in all three of these directions is obtained only if the impurity sits on a substitutional lattice site (see lattice model in Fig. 11.22).

One can conclude from these experimental results that ^{8}Li impurities preferentially occupy tetrahedral interstitial sites after implantation at 300 K but convert to substitutional lattice positions after annealing at 450 K.

11.2.2 Detection of Epitaxial Growth

Channeling occurs only if undisturbed channels or lattice planes exist. As we saw above, impurities on nonsubstitutional sites can interfere with channeling. Another channeling disturbance is the distortion which occurs at the interface between two different crystals. Such a situation occurs in heteroepitaxy of a thin layer of one material on a single crystal of another.

We return again to the cobalt silicide system discussed in Section 11.1.5. There it was shown how Rutherford backscattering detects the formation of CoSi and $CoSi_2$ phases. One important aspect was not considered in that section, i.e. the question of whether these phases are monocrystalline and, if so, what is their orientation with respect to the Si surface. This question can be answered by channeling experiments, as will be shown for the case of $CoSi_2$.

In Fig. 11.23 the cross section through the crystal structure of $CoSi_2$, assumed to be epitaxial on silicon, is shown schematically. However, one should expect that the crystal lattice of $CoSi_2$ is distorted compared to its normal structure. If the channeling effect is detectable in $CoSi_2$ it proves that $CoSi_2$ is monocrystalline. From the deviation of the $\langle 110 \rangle$ axis in $CoSi_2$ from that in Si, one can determine the distortion of $CoSi_2$. Both effects were found in the experiment (Wu et al., 1991).

On the right side of Fig. 11.23 two channeling spectra are shown; the top spectrum is for the $CoSi_2$ film and the lower spectrum is for the silicon

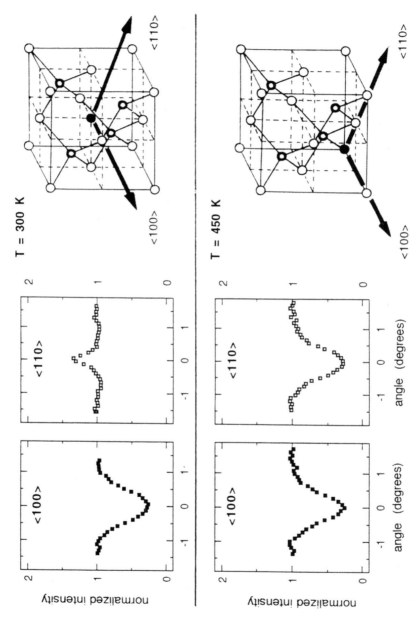

Fig. 11.22 Emission channeling for ^{8}Li in InP for two implantation temperatures (Wahl et al., 1992).

Nuclear Reaction Analysis (NRA)

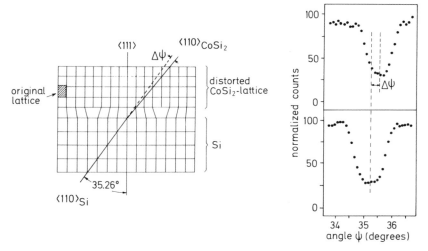

Fig. 11.23 Left side: Schematic representation of a CoSi$_2$ film which is grown epitaxially on Si. Right side: Channeling spectrum for the CoSi$_2$ film (top) and for the Si substrate (bottom). Adapted from Wu et al. (1991).

substrate. The two regions are distinguished by choosing the appropriate energy windows in the Rutherford backscattering spectrum. First, it is clear that channeling is indeed observed in CoSi$_2$, proving that the film is monocrystalline. Secondly, we note that the $\langle 110 \rangle$ axis of CoSi$_2$ is inclined by an angle of $\Delta\psi = 0.30°$ with respect to this axis in the Si substrate. One can conclude from this observation (Wu et al., 1991) that the CoSi$_2$ has a tetragonal distortion of

$$\varepsilon_T := \frac{d^\| - d_0^\|}{d_0^\|} - \frac{d^\perp - d_0^\perp}{d_0^\perp} = 1.1\% \qquad (11.44)$$

Here $d^\|$ and $d^\perp$ designate the separation between planes in epitaxial CoSi$_2$ parallel and perpendicular to the surface plane; and $d_0^\|$ and $d_0^\perp$ are these distances in the undistorted CoSi$_2$ crystal.

11.3 Nuclear Reaction Analysis (NRA)

The method of nuclear reaction analysis (NRA) identifies atoms in the sample by inducing nuclear reactions using accelerated particles and detecting the resulting radiation which is characteristic of the species. After normalizing to the number of incident particles, the intensity of this characteristic radiation is proportional to the concentration of this atomic species.

Two different principles are used in NRA for measuring concentration profiles:

(a) If the nuclear reaction has a resonance as a function of the energy of the incident particle, then the nuclear reaction takes place mainly at the depth where the projectile has slowed down to the resonant energy. By varying the incident energy, the resonance can be placed at different depths and thus the depth profile can be measured. This method is called nuclear resonance reaction analysis (NRRA).
(b) The second possibility for depth profiling uses the energy loss of the incident and outgoing particle to determine the depth where the reaction occurred. In this case the depth profile is calculated from the energy spectrum of the outgoing particles. A condition for using this method, however, is that the reaction can be clearly identified (e.g. by a high Q value for the reaction, so that no overlap with a competing reaction occurs) and that the cross section is known in the relevant energy range. For this reason, one selects an energy range in which the cross section varies smoothly with energy or, if possible, is constant.

Both methods, the resonant (NRRA) and nonresonant (NRA), are included in the general acronym NRA. An example of each of these two methods will be discussed below, and some reactions useful for materials analysis are summarized in a table.

11.3.1 Hydrogen Depth Profiling by the ^{15}N Method

Hydrogen is present, often as an unintentional impurity, in many materials and can have significant effects on materials properties. Examples are the saturation of bonds in amorphous silicon, hydrogen storage in metals, embrittlement of metals by hydrogen uptake, passivation of semiconductor dopants, etc. The possible importance of hydrogen as an energy carrier in a future hydrogen energy economy has contributed to the increasing interest of hydrogen in solids.

Hydrogen concentration as a function of depth can be determined reliably and quantitatively by application of the nuclear reaction

$$^{1}H(^{15}N, \alpha\gamma)^{12}C \tag{11.45}$$

which has a sharp resonance at the energy $E(^{15}N) = 6.40$ MeV. The cross section in the resonance is four orders of magnitude larger than outside the resonance, so that the reaction occurs predominantly at the resonant energy.

The characteristic radiation, the 4.43 MeV γ-ray emitted in the decay of the ^{12}C nucleus from the first excited state to its ground state, is measured by a NaI(Tl) detector (see Section 2.4). The detector should be as large as possible (e.g. 15 × 15 cm^2) in order to cover a large solid angle and obtain high detection efficiency for the relatively high γ-ray energy of 4.3 MeV. More recently, BGO (Bi–Ge–O) detectors have been used instead of NaI(Tl), because of the higher detection efficiency due to the heavy element Bi.

The principle of this ^{15}N analysis method is illustrated in Fig. 11.24. In this experiment the normalized number of γ-rays is measured as a function of

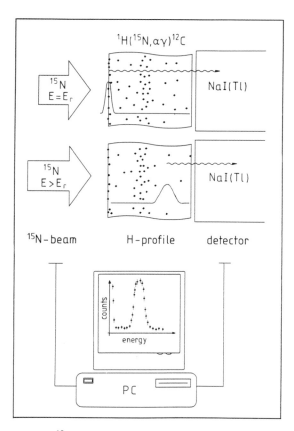

Fig. 11.24 Principle of the ^{15}N method. The hydrogen depth profile is measured by the nuclear reaction ^{1}H(^{15}N, αγ)^{12}C which has a sharp resonance at 6.40 MeV. The emitted 4.43 MeV γ-ray is detected in a NaI(Tl) detector. The depth profile is obtained by varying the incident energy. At $E(^{15}\text{N}) = E_γ$ the nuclear reaction takes place at the surface (top). At higher incident energies, the ^{15}N ion enters the sample and only when it has reached the resonance energy can the nuclear reaction take place (Boebel, 1991).

the incident ^{15}N energy (see picture on the PC monitor in the figure). This distribution gives the qualitative depth profile of the hydrogen.

If the host sample is homogeneous, the quantitative concentration profile can be obtained easily as a function of depth by rescaling the axes. For the x axis the relation between depth d in the sample and energy E of the incident ^{15}N ions is

$$d = (E - E_r)\left(\frac{dE}{dx}\right)^{-1} \qquad (11.46)$$

where E_r is the resonance energy (6.40 MeV) and dE/dx is the specific energy loss of ^{15}N ions in the material. In principle dE/dx depends on d or E and is also dependent on the local H concentration. However, both effects are very small and can be neglected. For most materials dE/dx depends very weakly (variations less than a few percent) on energy in the range from the resonance up to several MeV above the resonance, because this energy region is near the maximum energy loss region where dE/dx varies little with energy. The contribution of hydrogen to the energy loss is small because of the small mass of hydrogen and also is partially compensated by the lattice expansion induced by hydrogen, so that in effect a constant dE/dx can be used in Eq. (11.46).

The hydrogen concentration c is determined by rescaling the y axis from the measured number of γ-quanta N_γ normalized to the total integrated charge of the incident ion beam. The rescaling relation is

$$\frac{N_\gamma}{Q} \propto \frac{c}{dE/dx} \qquad (11.47)$$

where the proportionality constant depends only on the detector efficiency, the detector solid angle, and the resonance cross section. For a given geometry, these quantities are constant and equal for all samples, so that the apparatus can be calibrated by a measurement on a sample of known hydrogen concentration (e.g. a hydride of known stoichiometry). The concentration in the sample is then (in units of cm^{-3})

$$c = c_0 \frac{dE/dx}{(dE/dx)_0} \frac{N_\gamma}{N_{\gamma,0}} \frac{Q_0}{Q} \qquad (11.48)$$

The index 0 refers to the calibration material. The relatively uncertain dE/dx values are the main source of systematic errors in the absolute hydrogen concentration determinations by the ^{15}N analysis method.

Depth resolution of the ^{15}N method near the sample surface is determined by the energy spread of the beam, the width of the resonance, and vibrations of the H atoms, and is typically 5 nm. At larger depths, energy straggling of the ion beam is the main source of error in the depth determination. At a depth of 500 nm, the depth resolution is of order 10–15 nm.

Fig. 11.25 (Neitzert and Briere, 1989) shows the hydrogen depth profile of a p-i-n (p-type, intrinsic, n-type) solar cell made of amorphous silicon (a-Si:H). The a-Si:H layers were produced by a glow discharge in a cell filled with silane (SiH_4) and hydrogen. The molecules are broken up by the glow discharge and the products condensed on the substrate. At appropriate experimental conditions an amorphous silicon film is deposited on the substrate. Incorporation of hydrogen is critical, since hydrogen saturates broken silicon bonds which otherwise would produce states in the band gap. The p- and n-type material is obtained by adding B_2H_6 and PH_3 respectively to silane in concentrations of ca. 0.1%.

As can be seen from Fig. 11.25, the hydrogen content in this cell is of order 10 at.% and is somewhat higher in the doped layers than in the intrinsic region. This is a typical result for a-Si:H solar cells. However, since the hydrogen content in the a-Si:H layer depends strongly on the production conditions, other distributions are also possible. The behavior of hydrogen is of great importance for the long-time stability of solar cells made of amorphous silicon. Therefore it is a subject of current intensive interest.

For inhomogeneous samples—such as the multiple-layered systems discussed several times in this book—Eqs. (11.46)–(11.48) can be applied only piecewise within each homogeneous region because dE/dx is not constant. Difficulties arise at the interface of two materials. The value of dE/dx changes abruptly, but nuclear reactions take place on both sides of the interface because of the energy distribution of particles at that depth.

In principle a deconvolution of the measured spectrum through use of a

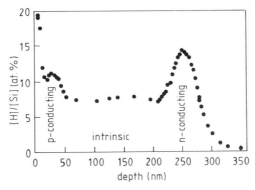

Fig. 11.25 Hydrogen depth profile of a p-i-n solar cell made of amorphous silicon. The hydrogen concentration near the surface is due to adsorbates (Neitzert and Briere, 1989).

resolution function is possible. However, in practice, this is relatively complicated, and a simulation procedure is usually applied. One assumes a certain hydrogen concentration, convolutes it with the resolution function, and adjusts the concentration until the experimental spectrum is fitted. An example is shown in Fig. 11.26 for a Au/Nb/Ti/Nb layer system (Boebel *et al.*, 1991). The sample was produced by evaporation of the metals onto an Al_2O_3 substrate in ultrahigh vacuum. Afterwards the sample was charged with hydrogen at 300 °C. The solid line shown in Fig. 11.26 was calculated under the assumption that the hydrogen concentration in titanium is 7 at.% and in the two Nb layers 0.4 at.%. In the Au layer, the hydrogen concentration is low, and in the Al_2O_3 substrate it is zero within the experimental uncertainties. The deviation of the data points from the theoretical curve in the region of the Nb/Ti interfaces indicates that the transition of the high H concentration in titanium to the low concentration in niobium is not abrupt but is somewhat smeared out.

One sees in Fig. 11.26 that H is strongly accumulated in the titanium layer. This is due to the fact that hydrogen is more strongly bound in Ti than in Nb. The critical quantity is the enthalpy of solution. The difference in these enthalpies determines the distribution of hydrogen among the two systems. For bulk samples, the solution enthalpies are known, but they may well be different for thin films. Differences between films and bulk might occur if dimensionality effects come into play in thin films or because expansion of the films on uptake of hydrogen is constrained by the substrate.

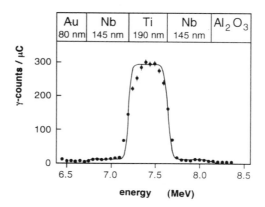

Fig. 11.26 Hydrogen depth profile of a Au/Nb/Ti/Nb layer system. The γ-ray count rate is normalized to the accumulated charge of the ion beam. The solid line is a fit assuming H concentrations of 7 at.% in Ti and 0.4 at.% in Nb. The H concentration in Au and the Al_2O_3 substrate is low (Boebel *et al.*, 1991).

11.3.2 Interdiffusion in Polymers Detected by the $^2H(^3He, {}^4He)^1H$ Nuclear Reaction

The interdiffusion of polymers can be studied by deuteration (^2H instead of ^1H) of polymers in a surface layer on a protonated polymer substrate. If the two substances interdiffuse (e.g. by annealing) the deuteron distribution, which is a step function at the beginning, is broadened. Since the deuterium concentration as a function of depth can be measured with the nuclear reaction, the interdiffusion of two polymers can be measured in this way. Deuterium is detected through the nuclear reaction

$$^2H(^3He, {}^4He)^1H \qquad Q = 18.352 \text{ MeV} \qquad (11.49)$$

Because of the large Q value, the reaction products have relatively large energies, and therefore can easily be separated from the background and from particles arising from competing reactions.

A schematic diagram of the experimental apparatus is shown in Fig. 11.27. The collimated ^3He beam of approximately 700 keV energy impinges on the sample and initiates nuclear reactions with the deuterium atoms present at some depth. The reaction products are detected at angle ϑ with respect to the incident beam using a semiconductor detector. In the present case, ^4He is measured. Discrimination against protons from the reaction is attained by using a detector just thick enough to stop ^4He particles. Since the protons have much smaller specific energy loss, the energy deposited in the detector by the protons is so small that they do not interfere with the ^4He in the energy spectrum. The slow but very abundant elastically scattered ^3He ions can be bent away from the detector by applying a magnetic field

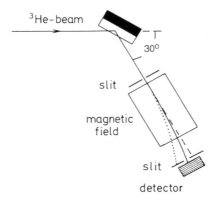

Fig. 11.27 A schematic illustration of the experimental set-up for the determination of deuterium depth profiles (Chaturvedi et al., 1990).

between the sample and detector. These low-momentum particles are deflected much more strongly than the fast ^{4}He ions. The ^{3}He ions can be separated out, and the detector loading reduced, through a system of apertures adjusted to the path of the ^{4}He ions.

The kinematics of the nuclear reaction are illustrated in Fig. 11.28. The index 3 and the corresponding reaction angle ϑ refer to the detected ^{4}He particle. The remaining notation is self-explanatory. In the nonrelativistic approximation, which is adequate for the present case, the energy E_3 of the ^{4}He particles after the reaction is (Mayer and Rimini, 1977)

$$E_3 = \frac{m_1 m_3 E_1}{(m_1 + m_2)(m_3 + m_4)}$$
$$\times \left[\cos \vartheta + \sqrt{\frac{m_2 m_4 E_1 + Q m_4 (m_1 + m_2)}{m_1 m_3 E_1} - \sin^2 \vartheta} \right]^2 \quad (11.50)$$

For $\vartheta = 30°$ and $E_1 = 700$ keV, $E_3 = E(^4\text{He}) = 5.83$ MeV. The energy of the accompanying (nondetected) ^{1}H particle is then $E_1 + Q - E_3 = 13.22$ MeV.

Depth information is provided by the energy of the detected ^{4}He particles. The highest energy is obtained for reactions which occur at the surface. For events occurring at greater depth, the energy loss of the ^{3}He projectile and of the emitted ^{4}He particle passing through the sample must be taken into consideration. The major contribution comes from the incident ^{3}He projectiles, since their dE/dx values at their relatively low energies are appreciably higher than those of the fast ejected ^{4}He particles.

In addition, there is a so-called energy magnification effect which arises because energy loss by the ^{3}He beam results in an even larger reduction of the ^{4}He particle energy. This arises from the kinematics of the nuclear reaction. The amplification factor, which is the derivative of E_3 with respect

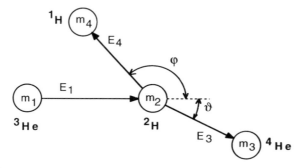

Fig. 11.28 Kinematics of the nuclear reaction.

to E_1 (see Eq. (11.50)) has the value $dE_3/dE_1 = 1.7$ for $E_1 = 700$ keV and $\vartheta = 30°$. Thus, if energy ΔE is lost by a ^{3}He incident particle, the ^{4}He particle energy is reduced by $1.7\Delta E$.

The kinematic quantities relevant to this nuclear reaction are summarized in Table 11.1 for two different ^{3}He energies. The specific energy loss in polystyrene for the incident ^{3}He and the outgoing ^{4}He particles are given in the last two columns.

As a concrete example of the application of the method, the interdiffusion of polystyrene will be discussed (Eiser et al., 1991). Styrene (phenyl ethylene) is a benzene derivative with the chemical formula C_6H_5—CH=CH$_2$. The corresponding polymer has the net composition $(C_8H_8)_n$ where n is the degree of polymerization.

For the investigation of the interdiffusion a 200 nm thick film of deuterated polystyrene (dPS) was deposited on a film of protonated polystyrene (hPS). The molecular weights were $M_W = 7.52 \times 10^5$ for dPS and $M_W = 6.6 \times 10^5$ for hPS, i.e. the degree of polymerization was slightly higher for dPS than for hPS.

Fig. 11.29 shows the measured dPS concentration as a function of depth for three different annealing temperatures. The graphs are ^{4}He spectra in which the y axis at the plateau is normalized to unity, and the x axis was transformed from energy loss to depth. The distribution before the annealing treatment (left side) was obtained at room temperature. One assumes that in this case the real dPS distribution is a step function and that the smearing of the distribution at the transition is due to the experimental resolution, which in the present case is approximately 50 nm.

After annealing the sample at 140 °C for different lengths of time, one observes a significant broadening of the interface region (Fig. 11.29, middle and right spectra). The broadening of the transition is caused by interdiffusion of dPS and hPS. If dPS and hPS are miscible (at 140 °C they are miscible), the new dPS distribution can be obtained by folding of the original step function with an error function. The width of the error function

Table 11.1 Kinematic quantities for the ^{2}H(^{3}He, ^{4}He)^{1}H nuclear reaction at $\vartheta(^4\text{He}) = 30°$. The quantity dE_3/dE_1 is the energy amplification factor. The specific energy loss in polystyrene for ^{3}He at the incident energy and for ^{4}He at the maximum outgoing energy is also given.

$E_1 = E(^3\text{He})$ (keV)	$E_3 = E(^4\text{He})$ (MeV)	$E_4 = E(^1\text{H})$ (MeV)	$\dfrac{dE_3}{dE_1}$	$\dfrac{dE(^3\text{He})}{dx}$ (eV nm^{-1})	$\dfrac{dE(^4\text{He})}{dx}$ (eV nm^{-1})
700	5.83	13.22	1.7	267	89
900	6.15	13.10	1.5	243	86

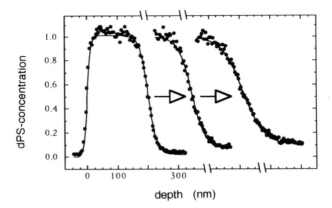

Fig. 11.29 Concentration distribution of deuterated polystyrene as a function of depth, directly after production (left side) and after annealing at 140 °C for 2×10^5 s (middle) and for 6×10^5 s (right side). The solid lines are fitted to a theoretical model in which the experimental resolution is assumed to be Gaussian and interdiffusion is described by an error function (Eiser et al., 1991).

gives the width of the interface w. Width w is associated with the diffusion coefficient D by the following relation,

$$w = \sqrt{2Dt} \qquad (11.51)$$

Fig. 11.30 shows that this relationship is well fulfilled. The slope of the line in Fig. 11.30 yields $D = 2.44 \times 10^{-21}$ m^2 s^{-1}.

Detailed investigations (Steiner et al., 1990) of the system dPS/hPS show that the interdiffusion depends on the concentration of dPS in hPS. The

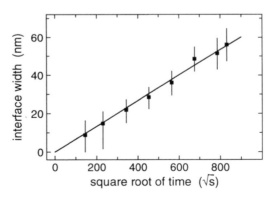

Fig. 11.30 Width of the interface layer as a function of the square root of the annealing time of the dPS/hPS sample at 140 °C (Eiser et al., 1991).

reason for this dependence is a repulsive interaction between dPS and hPS. At higher temperatures ($T > 140\,°C$) this repulsion is inconsequential, but at lower temperatures ($T_c = 60\,°C$) a miscibility gap occurs. This immiscibility influences the interdiffusion. We will not discuss this any further but refer the reader to the literature (Steiner *et al.*, 1990; Chaturvedi *et al.*, 1990).

11.3.3 A Few Nuclear Reactions Useful for NRA

The NRA method is used mainly for detection of light elements, since other methods (e.g. neutron activation analysis, Rutherford backscattering, PIXE) are less suitable for light elements. Some of the useful nuclear reactions for the NRA method are summarized in Table 11.2 (see also Mayer and Rimini, 1977). The resonance reactions can be identified by the appearance of a number for the resonance width Γ in the table. In these reactions the incident energy is varied and the energy dependence is scaled to depth. For the remaining reactions a fixed incident energy is chosen; information on depth is then obtained from the energy of the outgoing particles.

The precision of depth analysis obtained by the NRA method is different for the various reactions; it also depends somewhat on the depth and the sample under investigation. Values for the depth resolution in near-surface regions are typically 10 nm. At larger depths the resolution becomes worse, due to energy straggling of the ions. For resonance reactions, the depth to which the analysis can be performed depends on the energy at which the

Table 11.2 Selected nuclear reactions which are appropriate for the NRA method. For resonance reactions, the resonance energy E_r, the resonance width Γ, and the energy E_3 of the detected γ-ray are given. In this case the incident energy E_{inc} is varied. One starts the energy variation somewhat below the resonance in order to detect surface concentration. For nonresonance nuclear reactions, the Q value of the reaction, a typical incident energy E_{inc}, and a typical energy E_3 of the detected charged particle are given.

	Q (MeV)	E_{inc} (MeV)	E_r (MeV)	Γ (keV)	E_3 (MeV)
^{1}H(^{15}N, $\alpha\gamma$)^{12}C	—	—	6.385	1.8	4.43
^{2}H(^{3}He, α)^{1}H	18.352	0.7	—	—	5.8
^{7}Li(p, α)^{4}He	17.347	1.5	—	—	7.7
^{11}B(p, γ)^{12}C	—	—	0.163	5.2	4.43
^{12}C(d, p)^{13}C	2.722	1.2	—	—	3.1
^{15}N(p, $\alpha\gamma$)^{12}C	—	—	0.429	0.9	4.43
^{16}O(d, p)^{17}O	1.917	0.9	—	—	2.4
^{27}Al(p, γ)^{28}Si	—	—	0.992	0.1	10.78
^{30}Si(p, γ)^{31}P	—	—	0.620	0.07	7.90
^{31}P(p, γ)^{32}S	—	—	0.881	0.4	7.42

next resonance occurs. For nonresonant reactions the analysis depth is limited by background problems and by interference from other nuclear reactions. A typical value for the analysis depth is 1 μm.

In general, concentrations of 1% or less can be detected easily. In some special cases concentrations in the ppm region can be measured if adequate background suppression techniques are used.

After calibration of the apparatus using a standard, the NRA method gives absolute values for the concentrations. This is a major advantage over some other methods for which the results depend on the matrix. The NRA method determines the concentration of an isotope independent of its chemical bonding and also independent of the presence of other atoms. The precision of the concentration determination with NRA methods is of order 10%.

Appendix

A.1 Clebsch–Gordan Coefficients and 3j-Symbols

The coupling of two angular momenta j_1 and j_2 to the total angular momentum j_3 can be written

$$j_3 = j_1 + j_2 \quad (m_3 = m_1 + m_2) \tag{A.1}$$

The states of the entire system can be represented either by the products of $|j_1, m_1\rangle$ with $|j_2, m_2\rangle$ or by $|j_3, m_3\rangle$, i.e.

$$|j_1, m_1\rangle |j_2, m_2\rangle \quad \text{or} \quad |j_3, m_3\rangle \tag{A.2}$$

Thus there must be a linear relation between the two representations

$$|j_3, m_3\rangle = \sum_{\substack{m_1, m_2 \\ m_3 = m_1 + m_2}} (j_1, j_2, m_1, m_2 | j_3, m_3) |j_1, m_1\rangle |j_2, m_2\rangle \tag{A.3}$$

The expansion coefficients are called Clebsch–Gordan coefficients. Often one uses the Wigner $3j$-symbols since they possess higher symmetries. The $3j$-symbols are associated with the Clebsch–Gordan coefficients by the following relationship

$$\begin{pmatrix} j_1 & j_2 & j_3 \\ m_1 & m_2 & m_3 \end{pmatrix} = (-)^{j_1 - j_2 - m_3} \frac{1}{\sqrt{2j_3 + 1}} (j_1, j_2, m_1, m_2 | j_3, -m_3) \tag{A.4}$$

The Clebsch–Gordan coefficient for the maximum j_3, m_3 value can easily be given. For $j_3 = j_1 + j_2$ and $m_3 = m_1 + m_2$ there is only one term in the sum of Eq. (A.3). Because of the wave function normalization it must be true that

$$(j_1, j_2, j_1, j_2 \mid j_1 + j_2, j_1 + j_2) = 1 \tag{A.5}$$

The other Clebsch–Gordan coefficients can be obtained by using the raising and lowering operators j_+ and j_-. These are tabulated, e.g. in Rotenberg *et al.* (1959).

Some symmetry properties of the $3j$-symbols are (Lindner, 1984):

$$\begin{pmatrix} j_1 & j_2 & j_3 \\ m_1 & m_2 & m_3 \end{pmatrix} = \begin{pmatrix} j_2 & j_3 & j_1 \\ m_2 & m_3 & m_1 \end{pmatrix} = \begin{pmatrix} j_3 & j_1 & j_2 \\ m_3 & m_1 & m_2 \end{pmatrix} \tag{A.6}$$

$$(-)^{j_1+j_2+j_3} \begin{pmatrix} j_1 & j_2 & j_3 \\ m_1 & m_2 & m_3 \end{pmatrix} = \begin{pmatrix} j_2 & j_1 & j_3 \\ m_2 & m_1 & m_3 \end{pmatrix} \tag{A.7}$$

$$\begin{pmatrix} j_1 & j_2 & j_3 \\ m_1 & m_2 & m_3 \end{pmatrix} = (-)^{j_1+j_2+j_3} \begin{pmatrix} j_1 & j_2 & j_3 \\ -m_1 & -m_2 & -m_3 \end{pmatrix} \tag{A.8}$$

Special cases

$$\begin{pmatrix} j & j & 0 \\ m & -m & 0 \end{pmatrix} = (-)^{j-m} \frac{1}{\sqrt{2j+1}} \tag{A.9}$$

If $j_1 + j_2 + j_3$ is odd then

$$\begin{pmatrix} j_1 & j_2 & j_3 \\ 0 & 0 & 0 \end{pmatrix} = 0 \tag{A.10}$$

If $j_1 + j_2 + j_3$ is even then

$$\begin{pmatrix} j_1 & j_2 & j_3 \\ 0 & 0 & 0 \end{pmatrix} = (-)^p \sqrt{\frac{(2p - 2j_1)! \, (2p - 2j_2)! \, (2p - 2j_3)!}{(2p+1)!}}$$

$$\times \frac{p!}{(p - j_1)!(p - j_2)!(p - j_3)!} \tag{A.11}$$

with $2p := j_1 + j_2 + j_3$.

A.2 Spherical Tensors

An example of a second-rank tensor is the dyadic product of two vectors x and y. The tensor components are

$$T_{ik} = x_i y_k \quad (i, k = 1, 2, 3) \tag{A.12}$$

On rotation of the coordinate system, such a tensor usually transforms in a complicated way. There are combinations of tensor components which transform more simply, however, e.g. a scalar, vector, etc. In general, a tensor of second rank can be decomposed into irreducible tensors in the following way

$$T_{ik} = I_{ik} + A_{ik} + S_{ik} \tag{A.13}$$

with

$$I_{ik} = \frac{1}{3} \text{Tr}(T_{ik}) \delta_{ik}$$

$$A_{ik} = \frac{1}{2}(T_{ik} - T_{ki}) = -A_{ik} \tag{A.14}$$

$$S_{ik} = \frac{1}{2}(T_{ik} + T_{ki}) - \frac{1}{3} \text{Tr}(T_{ik}) \delta_{ik}$$

where δ is the Kronecker delta symbol. I_{ik} is a scalar (one component), A_{ik} a vector (three components) and S_{ik} a tensor (five components). The scalar can also be designated a tensor of zeroth rank and the vector a tensor of first rank.

Spherical tensors

We define the spherical components of a vector r as

$$r_{+1} = -\frac{1}{\sqrt{2}}(x + iy)$$

$$r_{-1} = \frac{1}{\sqrt{2}}(x - iy) \tag{A.15}$$

$$r_0 = z$$

or with the help of the spherical harmonics $Y_1^m(\theta, \phi)$

$$r_m = \sqrt{\frac{4\pi}{3}} r^1 Y_1^m(\theta, \phi) \tag{A.16}$$

Similar expressions can be formed for the components of other quantities which transform on rotation in the same way as a vector. Thus one can use the vector coupling method in order to build up spherical tensors of arbitrary rank. In general for spherical tensors $T(l, m)$ of lth rank one has

$$T(l, m) = \sum_{\substack{m_1, m_2 \\ m = m_1 + m_2}} (l_1, l_2, m_1, m_2 \mid l, m) T_1(l_1, m_1) T_2(l_2, m_2) \quad (A.17)$$

where $(l_1, l_2, m_1, m_2 \mid l, m)$ are Clebsch–Gordan coefficients.

EXAMPLES

(a) Scalar product of two vectors

$$T(0, 0) = \sum_{m_1 + m_2 = 0} (1, 1, m_1, m_2 \mid 0, 0) T_1(1, m_1) T_2(1, m_2) \quad (A.18)$$

$T(0, 0)$ is a scalar ($l = 0$), whereas $T_1(1, m_1)$ and $T_2(l, m_2)$ are vectors. The above equation is

$$\begin{aligned} T(0, 0) = {}& (1, 1, 1, -1 \mid 0, 0) T_1(1, 1) T_2(1, -1) \\ & + (1, 1, 0, 0 \mid 0, 0) T_1(1, 0) T_2(1, 0) \\ & + (1, 1, -1, 1 \mid 0, 0) T_1(1, -1) T_2(1, 1) \quad (A.19) \end{aligned}$$

(b) Vector product of two vectors

$$T(1, m) = \sum_{m_1 + m_2 = m} (1, 1, m_1, m_2 \mid 1, m) T_1(1, m_1) T_2(1, m_2)$$

$$(A.20)$$

The maximum component is

$$\begin{aligned} T(1, 1) = {}& (1, 1, 0, 1 \mid 1, 1) T_1(1, 0) T_2(1, 1) \\ & + (1, 1, 1, 0 \mid 1, 1) T_1(1, 1) T_2(1, 0) \quad (A.21) \end{aligned}$$

(c) Tensor product of two vectors

$$T(2, m) = \sum_{m_1 + m_2 = m} (1, 1, m_1, m_2 \mid 2, m) T_1(1, m_1) T_2(1, m_2)$$

$$(A.22)$$

Definition An irreducible spherical tensor operator is a set of $2l+1$ operators $T(l, m)$ which transform on rotation as the components of the spherical harmonics Y_l^m.

A.3 Wigner–Eckart Theorem

If $T(l, m)$ is a spherical tensor operator of lth rank, then the angular momentum coupling rules give

$$\langle I', M'|T(l, m)|I, M\rangle = (-)^{I'-M'} \begin{pmatrix} I' & l & I \\ -M' & m & M \end{pmatrix} \langle I'\|T(l)\|I\rangle \quad (A.23)$$

The quantity $\langle I'\|T(l)\|I\rangle$ is called the reduced matrix element. Its value does not depend on the z components M', M, and m. This theorem is very useful for comparing amplitudes of transitions within a given manifold of states. The relative transition strengths depend only on the $3j$-symbols.

Bibliography of Advanced Topics

Chapters 2 and 3

Frauenfelder, H. and Henley, E. M. (1974). *Subatomic Physics*. Prentice-Hall, Englewood Cliffs, N. J.
Jackson, J. D. (1962). *Classical Electrodynamics*. Wiley, New York.
Krane, K. S. and Halliday, D. (1987). *Introductory Nuclear Physics*. Wiley, New York.
Kopfermann, H. (1956). *Kernmomente*, 2nd edn. Akademische Verlagsgesllschaft, Frankfurt.

Chapter 4

Goldanskii, V. I. and Herber, R. H. (1968). *Chemical Applications of Mössbauer Spectroscopy*. Academic Press, New York.
Gonsor, U. (ed.) (1975). *Mössbauer Spectroscopy*. Topics in Applied Physics, vol. 5. Springer, Berlin.
Wegener, H. (1966). *Der Mössbauer-Effekt und seine Anwendungen in Physik und Chemie*. 2nd edn. BI Hochschultaschenbuch. Bibliographisches Institut, Mannheim.

Chapter 5

Christiansen, J. (ed.) (1983). *Hyperfine Interactions of Radioactive Nuclei*. Topics in Current Physics, vol. 31. Springer, Berlin.
Frauenfelder, H. and Steffen, R. M. (1965). In *Alpha-, Beta- and Gamma-Ray Spectroscopy*, vol. 2 (ed. K. Siegbahn). North-Holland, Amsterdam.
Rinneberg, H. H. (1979). Application of perturbed angular correlation to chemistry and related areas of solid state physics. *At. Energy Rev.* **17**, 477.

Chapter 6

Abragam, A. (1961). *The Principles of Nuclear Magnetism*. Clarendon, Oxford.
Shaw, D. (1976). *Fourier Transform N.M.R. Spectroscopy*. Elsevier, Amsterdam.
Slichter, C. P. (1978). *Principles of Magnetic Resonance*, 3rd edn. Springer Series in Solid State Science, vol. 1. Springer, Berlin.

Chapter 7

De Groot, S. R., Tolhoek, H. A. and Huiskamp, W. J. (1965). In *Alpha-, Beta- and Gamma-Ray Spectroscopy*, vol. 2 (ed. K. Siegbahn). North-Holland, Amsterdam.
Deutsch, B. I. and Vanneste, L. (eds) (1985). Nuclear orientation and nuclei far from stability. *Hyperfine Interact.* **22**.

Chapter 8

Schenck, A. (1985). *Muon Spin Rotation Spectroscopy*. Hilger, Bristol.
Seeger, A. (1978). In *Hydrogen in Metals I* (eds G. Alefeld and J. Völkl). Topics in Applied Physics, vol. 28. Spinger, Berlin.
Yamazaki, T. and Nagamine, K. (ed) (1984). Muon spin rotation and associated problems. *Hyperfine Interact.* **17–19**.

Chapter 9

Brandt, W. and Dupasquier, A. (eds) (1983). *Positron Solid State Physics*. Proc. Int. School Phys. 'Enrico Fermi', Course 83. North-Holland, Amsterdam.
Hautojärvi, P. (ed.) (1979). *Positrons in Solids*. Topics in Current Physics, vol. 12. Springer, Berlin.

Chapter 10

Bacon, G. E. (1975). *Neutron Diffraction*, 3rd edn. Clarendon, Oxford.
Bée, M. (1988). *Quasielastic Neutron Scattering*. Hilger, Bristol.
Lechner, R. E., Richter, D. and Riekel, C. (1983). *Neutron Scattering and Muon Spin Rotation*. Springer Tracts in Modern Physics, vol. 101. Springer, Heidelberg.

Chapter 11

Chu, W. K., Mayer, J. W. and Nicolet, M.-A. (1973). *Backscattering Spectrometry*. Academic Press, New York.
Feldman, L. C. and Mayer, J. W. (1986). *Fundamentals of Surface and Thin Film Analysis*. North-Holland, New York.
Mayer, J. W. and Rimini, E. (eds) (1977). *Ion Beam Handbook for Materials Analysis*. Academic Press, New York.
Morgan, D. V. (1973). *Channeling*. Wiley, London.

References

Abragam, A. (1961). *The Principles of Nuclear Magnetism*. Clarendon, Oxford.
Anderson, P. W. (1958). *Phys. Rev.* **109**, 1492.
Askill, J. (1970). *Tracer Diffusion Data for Metals, Alloys and Simple Oxides*. IFI/Plenum, New York.
Bacon, F., Barclay, J. A., Brewer, W. D., Shirley, D. A. and Templeton, J. E. (1972). *Phys. Rev.* B **5**, 2397.
Bacon, G. E. (1975). *Neutron Diffraction*, 3rd edn. Clarendon, Oxford.
Bardon, M., Norton, P., Peoples, J., Sachs, A. M. and Lee-Franzini, J. (1965). *Phys. Rev. Lett.* **14**, 449.
Bauminger, E. R., Froindlich, D., Nowik, I. and Ofer, S. (1973). *Phys. Rev. Lett.* **30**, 1053.
Beckurts, K. H. and Wirtz, K. (1964). *Neutron Physics*. Springer, Berlin.
Bée, M. (1988). *Quasielastic Neutron Scattering*. Hilger, Bristol.
Bloch, F. (1946). *Phys. Rev.* **70**, 460.
Bloch, F., Hansen, W. W. and Packard, M. (1946). *Phys. Rev.* **69**, 127.
Bloembergen, N. (1949). *Physica* **15**, 588.
Boebel, O. (1991). Dissertation, Universität Konstanz.
Boebel, O., Blässer, S., Steiger, J. and Weidinger, A. (1991). *Wissenschaft Fortschritt* **41**, 5.
Breman, M. and Boerma, D. O. (1991). Private communication.
Brewer, W. D., Shirley, D. A. and Templeton, J. E. (1968). *Phys. Lett.* **27A**, 81.
Butz, T. and Lerf, A. (1983). *Phys. Lett.* **97A**, 217.
Camani, M., Gygax, F. N., Rüegg, W., Schenck, A. and Schilling, H. (1977). *Phys. Rev. Lett.* **39**, 836.
Celio, M. and Meier, P. F. (1983). *Phys. Rev.* B **28**, 39.
Chaturvedi, U. K., Steiner, U., Zak, O., Krausch, G., Schatz, G. and Klein, J. (1990). *Appl. Phys. Lett.* **56**, 1228.
Christiansen, J., Heubes, P., Keitel, R., Klinger, W., Löffler, W., Sandner, W. and Witthuhn, W. (1976). *Z. Phys.* B **24**, 177.
Denison, A. B., Graf, H., Kündig, W. and Meier, P. F. (1979). *Helv. Phys. Acta* **52**, 460.
Dorner, B. (1982). *Coherent Inelastic Neutron Scattering in Lattice Dynamics*. Springer Tracts in Modern Physics, vol. 93. Springer, Berlin.
Doyama, M. and Hasiguti, R. R. (1973). *Cryst. Lattice Defects* **4**, 139.
Dybdal, K., Forster, J. S. and Rud, N. (1979). *Phys. Rev. Lett.* **43**, 1711.
Eiser, E., Steiner, U., Krausch, G., Schatz, G., Budkowski, A. and Klein, J. (1991). *Jahresbericht Nukleare Festkörperphysik*, p. 84. University of Konstanz.

Eldrup, M., Mogensen, O. E. and Evans, J. H. (1975). In *Fundamental Aspects of Radiation Damage in Metals* (eds M. T. Robinson and F. W. Young), p. 1127. Energy Research and Development Administration, Washington, DC.
Ellis, Y. A. (1973). *Nucl. Data Sheets* **9**, 319.
Ernst, H., Hagn, E., Zech, E. and Eska, G. (1979). *Phys. Rev.* B **19**, 4460.
Evans, J. S., Kashy, E., Naumann, R. A. and Petry, R. F. (1965). *Phys. Rev.* **138**, B9.
Feiock, F. D. and Johnson, W. R. (1969). *Phys. Rev.* **187**, 39.
Feldman, L. C. and Mayer, J. W. (1986). *Fundamentals of Surface and Thin Film Analysis*. North-Holland, New York.
Ferguson, A. J. (1965). *Angular Correlation Methods in Gamma-Ray Spectroscopy*. North-Holland, Amsterdam.
Fink, R., Wesche, R., Klas, T., Krausch, G., Platzer, R., Voigt, J., Wöhrmann, U. and Schatz, G. (1990). *Surf. Sci.* **225**, 331.
Flinn, P. A. (1978). In *Mössbauer Isomer Shifts* (eds G. K. Shenoy and F. E. Wagner), North-Holland, Amsterdam.
Frauenfelder, H. and Steffen, R. M. (1965). In *Alpha-, Beta- and Gamma-Ray Spectroscopy*, vol. 2 (ed. K. Siegbahn), p. 997. North-Holland, Amsterdam.
Frederikse, H. P. R. (1981). In *AIP 50th Anniversary, Physics Vade Mecum* (ed. H. L. Anderson), p. 288. American Institute of Physics, New York.
Gonsor, U. (1975). In *Mössbauer Spectroscopy* (ed. U. Gonsor). Topics in Applied Physics, vol. 5, p. 1. Springer, Berlin.
Grebinnik, V. G., Gurevich, I. I., Zhukov, V. A., Manych, A. P., Meleshko, E. A., Muratova, I. A., Nikol'skii, B. A., Selivanov, V. I. and Suetin, V. A. (1975). *Sov. Phys.–JETP* **41**, 777.
Grötzschel, R., Hentschel, E., Klabes, R., Kreissig, U., Neelmeijer, C., Assmann, W. and Behrisch, R. (1992). *Nucl. Instr. Meth.* B **63**, 77.
Heitjans, P., Körblein, A., Ackermann, H., Dubbers, D., Fujara, F. and Stöckmann, H.-J. (1985). *J. Phys.* F **15**, 41.
Hellwege, K.-H. (1976). *Einführung in die Festkörperphysik*. Springer, Berlin.
Herlach, D., Stoll, H., Trost, W., Metz, H., Jackman, T. E., Maier, K., Schaefer, H. E. and Seeger, A. (1977). *Appl. Phys.* **12**, 59.
Hofsäss, H. and Lindner, G. (1991). *Phys. Rep.* **201**, 121.
Holz, M. and Knüttel, B. (1982). *Phys. Blätter* **38**, 368.
Jackson, J. D. (1962). *Classical Electrodynamics*. Wiley, New York.
Jeffrey, G. A. (1981). In *AIP 50th Anniversary, Physics Vade Mecum* (ed. H. L. Anderson), p. 134. American Institute of Physics, New York.
Kamke, D. (1979). *Einführung in die Kernphysik*. Vieweg, Braunschweig.
Kanamori, J., Yoshida, H. K. and Terakura, K. (1981). *Hyperfine Interact.* **9**, 363.
Katayama, I., Morinobu, S. and Ikegami, H. (1975). *Hyperfine Interact.* **1**, 113.
Kehr, K. W. (1984). *Hyperfine Interact.* **17–19**, 63.
Klein, E. (1977). *Hyperfine Interact.* **3**, 389.
Klein, O. and Nishina, Y. (1929). *Z. Phys.* **52**, 853.
Knight, W. D. (1949). *Phys. Rev.* **76**, 1259.
Kocher, D. C. (1974). *Nucl. Data Sheets* **11**, 279.
Kopfermann, H. (1956). *Kernmomente*, 2nd edn. Akademische Verlagsgesellschaft, Frankfurt.
Korecki, J. and Gradmann, U. (1985). *Phys. Rev. Lett.* **55**, 2491.
Korecki, J. and Gradmann, U. (1986). *Europhys. Lett.* **2**, 651.
Lamb, W. E. (1941). *Phys. Rev.* **60**, 817.
Lang, G., de Benedetti, S. and Smoluchowski, R. (1955). *Phys. Rev.* **99**, 596.

Lechner, R. E. (1983). In *Mass Transport in Solids* (eds F. Beniere and C. R. A. Catlow). Plenum, New York.
Lederer, C. M. and Shirley, V. S. (eds) (1978). *Table of Isotopes*, 7th edn. Wiley, New York.
Lindgren, B. (1986). *Phys. Rev.* B **34**, 648.
Lindgren, B. (1990). *Europhys. Lett.* **11**, 555.
Lindner, A. (1984). *Drehimpulse in der Quantenmechanik.* Teubner Studienbücher. Teubner, Stuttgart.
Lounasmaa, O. V. (1974). *Experimental Principles and Methods Below 1 K.* Academic Press, London.
Maier, K. (1984). *Hyperfine Interact.* **17–19**, 3.
Matthias, E., Schneider, W. and Steffen, R. M. (1962). *Phys. Rev.* **125**, 261.
Mayer, J. W. and Rimini, E. (eds) (1977). *Ion Beam Handbook for Materials Analysis.* Academic Press, New York.
Mayer-Kuckuk, T. (1984). *Kernphysik*, 4th edn. Teubner Studienbücher, Stuttgart.
Metag, V., Habs, D. and Specht, H. J. (1980). *Phys. Rep.* **65**, 1.
Metzner, H., Sielemann, R., Klaumünzer, S. and Hunger, H. (1987). *Mater. Sci. Forum* **15–18**, 1063.
Metzner, H., Sielemann, R., Butt, R. and Klaumünzer, S. (1984). *Phys. Rev. Lett.* **53**, 290.
Möslang, A., Graf, H., Balzer, G., Recknagel, E., Weidinger, A., Wichert, Th. and Grynszpan, R. I. (1983). *Phys. Rev.* B **27**, 2674.
Mössbauer, R. L. (1958). *Z. Phys.* **151**, 124.
Neitzert, H. C. and Briere, M. (1989). *J. Non-Cryst. Solids* **115**, 75.
Nicklow, R. M., Gilat, G., Smith, H. G., Raubenheimer, L. J. and Wilkinson, M. K. (1967). *Phys. Rev.* **164**, 922.
Patterson, B. D., Kündig, W., Meier, P. F., Waldner, F., Graf, H., Recknagel, E., Weidinger, A., Wichert, Th. and Hintermann, A. (1978). *Phys. Rev. Lett.* **40**, 1347.
Potzel, W., Forster, A. and Kalvius, G. M. (1978). *Phys. Lett.* **67A**, 421.
Purcell, E. M., Torrey, H. C. and Pound, R. V. (1946). *Phys. Rev.* **69**, 37.
Quitmann, D., Jaklevic, J. M. and Shirley, D. A. (1969). *Phys. Lett.* **30B**, 329.
Raghavan, P., Senba, M. and Raghavan, R. S. (1978). *Hyperfine Interact.* **4**, 330.
Raman, S. and Kim, H. J. (1971). *Nucl. Data Sheets* **6**, 39.
Recknagel, E., Schatz, G. and Wichert, Th. (1983). In *Hyperfine Interactions of Radioactive Nuclei* (ed. J. Christiansen). Topics in Current Physics, vol. 31, p. 133. Springer, Berlin.
Richter, D. and Shapiro, S. M. (1980). *Phys. Rev.* B **22**, 599.
Riegel, D., Bräuer, N., Focke, B., Lehmann, B. and Nishiyama, K. (1972). *Phys. Lett.* **41A**, 459.
Rotenberg, M., Bivins, R., Metropolis, N. and Wooten, J. K. Jr (1959). *The 3-j and 6-j Symbols.* MIT Press, Cambridge, MA.
Rowe, J. M., Rush, J. J., de Graaf, L. A. and Ferguson, G. A. (1972). *Phys. Rev. Lett.* **29**, 1250.
Sakamoto, M., Brockhouse, B. N., Johnson, R. G. and Pope, N. K. (1962). *J. Phys. Soc. Japan*, **17**, Suppl. B-II, 370.
Schatz, G., Ding, X. L., Fink, D., Krausch, G., Luckscheiter, B., Platzer, R., Voigt, J., Wöhrmann, U. and Wesche, R. (1990). *Hyperfine Interact.* **60**, 975.
Seeger, A. (1978). In *Hydrogen in Metals I* (eds G. Alefeld and J. Völkl). Topics in Applied Physics, vol. 28, 349. Springer, Berlin.
Shaw, D. (1976). *Fourier Transform N.M.R. Spectroscopy.* Elsevier, Amsterdam.

Shull, C. G., Wollan, E. O., Morton, G. A. and Davidson, W. L. (1948). *Phys. Rev.* **73**, 842.
Sigle, W., Carstanjen, H.-D., Flik, G., Herlach, D., Jünemann, G., Maier, K., Rempp, H., Seeger, A., Abela, R., Anderson, D. and Glasow, P. (1984). *Nucl. Instr. Meth.* B **2**, 1.
Speidel, K.-H. (1985). *Hyperfine Interact.* **22**, 305.
Stachel, M. (1982). Doktorarbeit, Universität Konstanz.
Steffen, R. M. (1956). *Phys. Rev.* **103**, 116.
Steiner, U., Krausch, G., Schatz, G. and Klein, J. (1990). *Phys. Rev. Lett.* **64**, 1119.
Stevens, J. G. and Stevens, V. E. (eds) (1977). *Mössbauer Effect Data Index 1976*. IFI/Plenum, New York.
Stone, N. J. and Hamilton, W. D. (1984). *Hyperfine Interact.* **10**, 1219.
Sze, S. M. (1985). *Semiconductor Devices—Physics and Technology*. Wiley, New York.
Tranquada, J. M., Moudden, A. H., Goldman, A. I., Zolliker, P., Cox, D. E., Shirane, G., Sinha, S. K., Vaknin, D., Johnston, D. C., Alvarez, M. S., Jacobson, A. J., Lewandowski, J. T. and Newsam, J. M. (1988). *Phys. Rev.* B **38**, 2477.
Vajda, S., Sprouse, G. D., Rafailovich, M. H. and Noé J. W. (1981). *Phys. Rev. Lett.* **47**, 1230.
Van Hove, L. (1954). *Phys. Rev.* **95**, 249.
Vantomme, A. (1991). Dissertation, Universität Leuven.
Vianden, R. (1983). *Hyperfine Interact.* **15/16**, 1081.
Wahl, U., Hofsäss, H., Jahn, S. G., Winter, S. and Recknagel, E. (1992). *Nucl. Instr. Meth.* B **64**, 221.
Wegener, H. (1966). *Der Mössbauer-Effekt und seine Anwendungen in Physik und Chemie*, 2nd edn. BI Hochschultaschenbuch, Mannheim.
Winnacker, A., Ackermann, H., Dubbers, D., Mertens, J. and von Blanckenhagen, P. (1971). *Z. Phys.* **244**, 289.
Witthuhn, W. and Engel, W. (1983). In *Hyperfine Interactions of Radioactive Nuclei* (ed. J. Christiansen). Topics in Current Physics, vol. 31, p. 205. Springer, Berlin.
Wu, C. S., Ambler, E., Hayward, R. W., Hoppes, D. D. and Hudson, R. P. (1957). *Phys. Rev.* **105**, 1413.
Wu, M. F., Vantomme, A., Langouche, G., Vanderstraeten, H. and Bruynseraede, Y. (1991). *Nucl. Instr. Meth.* B **54**, 444.
Ziegler, J. F. (1977). *The Stopping and Ranges of Ions in Matter*, vol. 4. Pergamon, New York.

Index

Absorption
  coefficient for Ge   17
  coefficient for NaI   17
  signal   109, 113
  recoilless   33
Alignment of nuclear spins   65, 139
Anderson localization   173
Angular correlation   63
  rotation of   77, 97
  for ^{60}Co   66
  for the 0-1-0 γ-γ-cascade   65
Angular distribution   12–14, 154–155
  anisotropic   63, 135, 140
  pattern   12–14
Anisotropy
  coefficient A_k 71
  of γ-radiation   63, 140
  of the muon decay   152, 155
Annealing (isochronal)   91–93, 192
Annihilation
  γ-ray   182
  of positrons   181
Arizona muons   153
Arrhenius behavior (curve, plot)   128, 169–171
Asymmetry parameter   29, 32
Atomic defects (in metals)   89
Autocorrelation function   213–214, 216

BaF$_2$ scintillator   18
Band state   171–172
Band-pass filter   114
Barn   7
Barrier hight   169
β-NMR   135–136
Bethe–Bloch formula   228–229

B field
  internal (local)   58–59, 96, 157, 162
  transient   98–99
  at surfaces   59–60
Bloch   101
  method   111
  equations   104, 106, 108
Blocking (effect)   247–249
Boltzmann distribution   101, 139–141
Boron trifluoride (BF$_3$) detector   199
Bose–Einstein distribution function   40
Bragg
  condition   198, 204
  line   204
  reflection   198, 206
Bragg–Kleeman rule   229
Breit–Rabi diagram   175, 178

Charged particle activation analysis (CPAA)   223
Channeling   223, 243–251
Chemical
  shift   118–120
  valence   50
Clebsch–Gordan coefficients   263–264
Cobalt silicide formation (system)   238, 249
Coincidence count rate   81
  random   82
  real   82
Compton effect   15–18
Conduction electron
  density   188
  polarized   123
Constant fraction discriminator   84–85
Contact interaction   126

276 Index

Continuous-wave method   110–112
Conversion coefficient   44–45
Conversion electron Mössbauer spectroscopy (CEMS)   49–50, 59
Coordinate system, rotating   106–109, 111
Core polarization   7, 9
Correlatation
  function   164, 213
  time   164, 129–130
Cross section
  for Compton effect   16
  for neutron scattering   201, 205
  for photoeffect   15
  for Rutherford scattering   226–227
Crystal monochromator   198
Curie law   103

Debye
  model   41–42
  temperature   41
Debye–Waller factor   36–43, 206–207
Decay
  of the muon   152–153
  of the neutron   195–196
  of the pion   151–152
  of the positron   181
  of ^{111}Sn   89
Decay scheme
  of ^{57}Co   44
  of ^{60}Co   67
  of ^{64}Cu   184
  of ^{67}Ga   45
  of ^{151}Gd   45
  of ^{181}Hf   79
  of ^{111}In   79
  of ^{177}Lu   144
  of ^{54}Mn   144
  of ^{22}Na   184
  of ^{100}Pd   80
  of ^{151}Sm   45
  of ^{119}Sn   45
Deformation parameter   9
Density of states
  of free electrons   124
  of phonons   41
Detector telescope for charged particles   234
Deuterium depth profile   257
Diamagnetic shift   103, 119–122
Diffusion   128, 162, 164–173, 259–260
  classical   169

coefficient   166, 168–169, 260
models   169–173
Dilution refrigerator   141–143
Dipole
  field   126, 163
  interaction, electric   26
  interaction, magnetic   21, 56
  moment, magnetic   3
  radiation   14
Dipole–dipole interaction   125, 162
Dirac values   5–6
Dispersion
  signal   109, 113
  relation   219–220
Doppler effect (shift)   35, 46, 48, 60–62, 182, 184–185
  quadratic   60–62
Dulong–Petit law   62

Effective charge   9
Einstein model   37, 40–41, 221
Elastic recoil detection analysis (ERDA)   240
Electric interaction   25, 27
Electron–positron pair   182
Elementary excitation   219
Emission
  channeling   247
  recoilless   33
Energy (level) splitting   21–22, 32, 53–54, 56–57
Energy loss
  in matter   227–229
  spectrum   221
Epitaxial growth   249, 251
Equilibrium, thermal   101
E-radiation   11
ERDA method   240
Eu compounds   52

Fast/slow principle   81
Field gradient, electric   29, 53–54, 77, 86
  axially symmetric   30
  from lattice ions   88
  from conduction electrons   88
  in cadmium   86–87
  in non-cubic metals   86–87
  on silver surfaces   92–93
  temperature dependence of   88–89

Fermi
  contact field   158, 161
  contact interaction   126, 174
  distribution function   124, 132
  energy   124, 188
  golden rule   132
  momentum   186
  surface   187, 189
  temperature   124, 188
  vector   188
Ferromagnetic (materials)   57–60, 94–95, 98–99, 145, 157–161
Fission isomers   9
Four-detector appparatus   83
Fourier transformation (spectrum)   117, 156–157, 180
Fractional distillation of ^{3}He   142
Free electron gas   124, 186–187
Frenkel pair   90
Friedel oscillations   60

γ-γ angular correlation   63, 182
  integral perturbed   95
  perturbed   63, 71
  unperturbed   63
γ-angular distribution   94
γ-decay   10
γ-NMR   135
γ-radiation   15
Geometrical structure factor   206–207
  for NaH   208
Geometry, perpendicular   75, 96
Germanium detector   19–20
g-factor   4
  in the single-particle (nucleon) model   4–7
Gyromagnetic ratio   4, 102, 153

^{3}He–^{4}He dilution refrigerator   141–143
^{3}He (neutron) detector   199
Hydrogen
  depth profile   252–255
  isotopes   171
Hyperfine field   95, 146
Hyperfine interaction   21, 126–127
  constant   127, 174
  in muonium   174–178
  sign of the   146

Implantation   78, 145
Induction decay, free-   116–118
Ion beam analysis   223

IPAC method   95
Isomer shift   27, 48–53
Isotope shift   27

Kinematic factor   224–226, 241
Kinematics for ^{2}H(^{3}He,α)^{1}H   258
Klein–Nishina formula   16
Knight shift   123, 127, 134
  of some metals   127
Korringa
  relation   131, 134
  scattering   148

Landé formula   4–5
Larmor frequency   23, 76, 154
Lattice
  site determination of impurities in crystals   247
  vector, reciprocal   204, 219
Leading-edge discriminator   83–84
Level
  isomeric   78
  splitting   21–22
Lifetime measurement (for positrons)   185–186, 189, 191–193
Line intensity (at the Mössbauer effect)   55–56, 59
Linewidth   110
  motional narrowing   130, 162
  natural   34–35
  quasielastic   218
Local mode   221
Lock-in amplifier   113–115
Lorentz
  distribution   34, 215
  field   158, 160–161
  sphere   157–158

Magnetic fields
  internal (local)   57–58, 94–97, 139–140, 157
  at surfaces   59–60
  transient   98–99
Magnetic interaction   21–22, 56, 74, 101
Magnetic structure of YBa$_2$Cu$_3$O$_6$ 211–212
Magnetization   101–109
  free motion equation   104
Matrix element, reduced   267
Maxwell
  equations   10
  velocity distribution   36

278 Index

Miller indices   204
Moderator temperature   197
Mössbauer
  apparatus   46
  drive   46–47
  effect   33
  line   36
  sources   43
Monochromator   198
Monopole term   27–28, 48
Motional narrowing   130, 162
  formula   164
M-radiation   11
Multipole
  fields   10
  operator   55, 64
  transitions   11
Multi-count mode   47
Muon beams   153
Muon diffusion   162
  in copper   165
  in iron   167–168
Muonium (in semiconductors)   174
  anomalous   174
  eigenstates   177
  energy eigenvalues   177
  normal   174
Muon
  polarization   151, 153
  trapping (by lattice defects)   166
  spin rotation (μSR)   150
Muons
  fast   154
  properties of   153
  static   162–163

NaI(Tl) scintillator   18
Neutrino recoil   89–90
Neutron(s)
  capture   78
  diffraction   202
  monochromatic   198
  polarized   135
  properties of   195–196
Neutron scattering   195, 201
  by condensed matter   202
  coherent   205–206
  elastic   195, 207
  incoherent   205
  inelastic   195, 219
  quasielastic   195, 212–213
90° pulse   107, 116

^{15}N method   252–253
NMR   101
  classical treatment of   104
  continuous wave   110–111
  experimental setup   103, 110
  nuclei   111
  on oriented nuclei   145
  pulsed   110, 116
  spin-echo method   118
  stationary   110
  with radioactive nuclei   134
NMR/NO   145
Nuclear
  dipole moment   3
  charge distribution   7–9
  magnetic resonance (NMR)   101
  magnetization   101
  magneton   4
Nuclear moments
  of ^{111}Cd   79
  of ^{151}Eu   45
  of ^{57}Fe   44
  of ^{177}Lu   144
  of ^{54}Mn   144
  of ^{100}Rh   80
  of ^{119}Sn   45
  of ^{181}Ta   79
  of ^{67}Zn   45
Nuclear
  orientation (NO)   139
  radius, mean square   27, 48
  reaction analysis (NRA)   223, 251
  reactions for the NRA method   261
  resonance   101
Nuclear spin
  polarisation   102–103, 139
  precession   22

Octahedral sites   159, 216, 218
Ortho-positronium   181

PAC (perturbed angular
    correlation)   63
  apparatus   81
  method, time differential   81
  method, integral   95
  sources   78
Paramagnetic shift (enhancement)   103, 119, 120, 123
Para-positronium   181
Parity   10, 11
  violation   139

Particle accelerators 229
Particle detectors 231
Paschen–Back region 132, 176
Pauli susceptibility 124
Perturbation factor (in PAC) 72–75
Phase
  diagram for ^{3}He-^{4}He 142
  difference (in neutron scattering) 202
  relaxation 105
Phonon
  density of states 41
  dispersion curve (branch) 220
Photoeffect 15
Photon 10
PIXE method 223
Plastic scintillator 19, 156
Point charge model 88
Poisson equation 26
Polaron 170
Polymer interdiffusion 257
Population probability 66–68
Positron 181
  annihilation 181
  lifetime (mean) 183, 185–186, 189
  sources 183–184
Positronium 181
Poynting vector 12
Precession 22–23, 75, 105
  in muonium 178
Proportional counter 47–48, 199
Purcell 101
  bridge 112
  method 111

Quadrupole
  coupling constant 86
  frequency 31, 76
  interaction, electric 27, 30, 53, 76, 146
Quadrupole moment
  elecric 7
  in single-particle model 7–9
  for charged ellipsoid 9
Quadrupole
  nuclear orientation 146–147
  operator 7
  radiation 13–14

Radiation
  angular pattern 12
  damage in metals 89, 192

Radiofrequency (RF)
  field 101, 104, 106
  waves 103
Rate equation 190, 216
Relaxation 105
  longitudinal 105
  spin-lattice 105
  spin–spin 105
  transverse 105
Residence time, mean 129, 166, 216
Resonance absorption
  of radiofrequency waves 103
  of γ-radiation 49
Rotating
  coordinate system 106
  magnetic field 104
Rutherford backscattering (RBS) 223, 226, 235–239

Scalar product 266
Scattering
  angle 203
  coherent 205
  cross section 203, 205, 226–227
Scattering function 213
  for H in palladium 217
  for water 215
  incoherent 213
Scattering
  geometry 46
  incoherent 205
  length (mean) 201–202, 204
  vector 203
Schmidt values 5, 6
Scintillation detector 18
Self-diffusion 128–129
  coefficient 214
  in Li 136, 137
Self-trapping 170, 191
Semiconductor dedector 19, 231
Slow passage (NMR) 108
Spherical tensor 265, 267
Spin-echo method 118
  exchange scattering 148
Spin-flip scattering 131
Spin–lattice relaxation 105
  at low temperatures 146
  in metals 131
Spin–spin relaxation 105
Spin–wave 60
Square displacement, average
  (mean) 38–41, 89, 206, 245

Static interaction 73–74
Sternheimer factor 88
Surface
  (adsorbate) sites 92–94
  barrier detector 231, 234
  magnetic hyperfine field 59
  muons 153
Stopping power 228–229
Structure factor 206–208
Superconductor, oxide 209
Superstructure (magnetic) 212

$T^{3/2}$ temperature dependence
  of the electric field gradient 89
  of the magnetic hyperfine field 60
Tensor
  electric field gradient 29
  nuclear electric quadrupole 29
  operator 267
  product 266
  spherical 265
Tetrahedral site 159–160, 218
Three-axis spectrometer 200
$3j$-symbol 55, 64, 66, 263–264
Time measurement, electronic apparatus for 83
Time-evolution (-development) operator 22, 71, 74
Time-of-flight spectrometer 200
Time-to-amplitude converter (TAC) 84–85
Time-to-digital converter (TDC) 84
Tin compounds 50
Tracer methods 2
Transient (magnetic) field 97–99
Transition
  amplitude 69–70
  probability 64
Transmission geometry 46

Trapping
  by lattice defects 166
  model for positrons 191
Tunneling
  coherent 171
  incoherent 170–171
  matrix element 170
  probability 170

Unit cell
  of NaH 208
  of $YBa_2Cu_3O_7$ 211

Vacancy
  concentration 189
  formation energy 189, 193
  in copper 90
  in gold 190–191
  in molydenium 192
Valence fluctuation 52–53
Van de Graaff accelarator 230
Van Hove autocorrelation function 213
Vector
  spherical harmonics 10
  potential 120
  product 266

Wavelength
  of the neutron 195–196
  of X-rays 196
Wigner–Eckart theorem 267
Wigner $3j$-symbol 263
Wu experiment 139

Zeeman
  effect 22
  energy 175
  region 175
Zero-point motion 40